Bioenergy and Environmental Biotechnology for Sustainable Development

Multidisciplinary Applications and Advances in Biotechnology: Contributions from the Biotechnology Society of Nigeria Working Groups

Series Editors:

Benjamin Ewa Ubi

Department of Biotechnology / Biotechnology Research & Development Centre, Ebonyi State University Abakaliki, Nigeria

Abdulrazak B. Ibrahim

Department of Biochemistry, Ahmadu Bello University, Zaria, Nigeria / Forum for Agricultural Research in Africa (FARA), Accra, Ghana

Akinola Rasheed Popoola

Department of Plant Pathology / Biotechnology Centre, Federal University of Agriculture, Abeokuta, Nigeria

Emeka Godfrey Nwoba

Algae Research & Development Centre, Murdoch University, Western Australia

James Chukwuma Ogbonna

Department of Microbiology, University of Nigeria, Nsukka, Nigeria

Charles Oluwaseun Adetunji

Edo State University Uzairue, Nigeria

Nwadiuto (Diuto) Esiobu

Florida Atlantic University, Boca Raton, FL., USA

Fermentation and Algal Biotechnologies for the Food, Beverage and Other Bioproduct Industries

James Chukwuma Ogbonna, Sylvia Uzochukwu, Emeka Godfrey Nwoba, Charles Oluwaseun Adetunji, Nwadiuto (Diuto) Esiobu, Abdulrazak Ibrahim, and Benjamin Ewa Ubi

Agricultural Biotechnology, Biodiversity and Bioresources Conservation and Utilization

Olawole Obembe, Emmanuel Olufemi Ekundayo, Arinze Stanley Okoli, Abubakar Gidado, Charles Oluwaseun Adetunji, Abdulrazak Ibrahim, and Benjamin Ewa Ubi

Medical Biotechnology, Biopharmaceutics, Forensic Science and Bioinformatics

Hajia Mairo Inuwa, Ifeoma Maureen Ezeonu, Charles Oluwaseun Adetunji, Abubakar Gidado, Emmanuel Olufemi Ekundayo, Abdulrazak Ibrahim, and Benjamin Ewa Ubi

Biosafety and Bioethics in Biotechnology: Policy, Advocacy, and Capacity Building

Sylvia Uzochukwu, Okoli Arinze Stanley Okoli, Nwadiuto (Diuto) Esiobu, Emeka Godfrey Nwoba, Christpeace Nwagbo Ezebuiro, Charles Oluwaseun Adetunji, Abdulrazak Ibrahim, and Benjamin Ewa Ubi

Microbiomes and Emerging Applications

Nwadiuto (Diuto) Esiobu, James Ogbonna, Charles Oluwaseun Adetunji, Olawole Obembe, Ifeoma Maureen Ezeonu, Abdulrazak Ibrahim, and Benjamin Ewa Ubi

Bioenergy and Environmental Biotechnology for Sustainable Development

Akinola Rasheed Popoola, Emeka Godfrey Nwoba, James Chukwuma Ogbonna, Charles Oluwaseun Adetunji, Nwadiuto (Diuto) Esiobu, Abdulrazak Ibrahim, and Benjamin Ewa Ubi

For more information about this series, please visit: https://www.routledge.com/Multidisciplinary-Applications-and-Advances-in-Biotechnology/book-series/MAABNIG

Bioenergy and Environmental Biotechnology for Sustainable Development

Edited by
Akinola Rasheed Popoola, Emeka Godfrey Nwoba,
James Chukwuma Ogbonna,
Charles Oluwaseun Adetunji,
Nwadiuto (Diuto) Esiobu, Abdulrazak B. Ibrahim,
and Benjamin Ewa Ubi

CRC Press
Taylor & Francis Group
Boca Raton London New York

CRC Press is an imprint of the
Taylor & Francis Group, an informa business

First edition published 2022
by CRC Press
6000 Broken Sound Parkway NW, Suite 300, Boca Raton, FL 33487-2742

and by CRC Press
4 Park Square, Milton Park, Abingdon, Oxon, OX14 4RN

CRC Press is an imprint of Taylor & Francis Group, LLC

Library of Congress Cataloging-in-Publication Data
Names: Ubi, Benjamin Ewa, 1968- editor. | Biotechnology Society of Nigeria, issuing body.
Title: Multidisciplinary applications and advances in biotechnology : contributions from the Biotechnology Society of Nigeria working groups, six-volume set / edited by Benjamin Ewa Ubi [and fourteen others]. Description: First edition. | Boca Raton : CRC Press, 2022. | Includes bibliographical references and index. |
Contents: Volume 1. Fermentation and Algal Biotechnologies for the Food, Beverage and Other Bioproduct Industries—Volume 2. Agricultural biotechnology, biodiversity and bio-resources conservation and utilization—Volume 3. Medical Biotechnology, Biopharmaceutics, Forensic Science and Bioinformatics—Volume 4. Biosafety and Bioethics in Biotechnology—Volume 5. Microbiomes and Emerging Applications—Volume 6. Bioenergy and environmental biotechnology for sustainable development, |
Summary: "This six volume book set examines a range of topics and applications related to biotechnology. Volumes include fermentation and algal biotechnologies; agricultural biotechnology; medical biotechnology, biopharmaceutics; biosafety, bioethics, biotechnology policy; microbiomes; bioenergy and environmental biotechnology for sustainable development"—Provided by publisher. Identifiers: LCCN 2021047039 (print) | LCCN 2021047040 (ebook) | ISBN 9780367766948 (v. 1 ; hardback) | ISBN 9781032013763 (v. 1 ; paperback) | ISBN 9780367766955 (v. 2 ; hardback) | ISBN 9781032014982 (v. 2 ; paperback) | ISBN 9780367766962 (v. 3 ; hardback) | ISBN 9781032015026 (v. 3 ; paperback) | ISBN 9780367767020 (v. 4 ; hardback) | ISBN 9781032015781 (v. 4 ; paperback) | ISBN 9780367767044 (v. 5 ; hardback) | ISBN 9781032018331 (v. 5 ; paperback) | ISBN 9780367767051 (v. 6 ; hardback) | ISBN 9781032018416 (v. 6 ; paperback) | ISBN 9781003178378 (v. 1 ; ebook) | ISBN 9781003178880 (v. 2 ; ebook) | ISBN 9781003178903 (v. 3 ; ebook) | ISBN 9781003179177 (v. 4 ; ebook) | ISBN 9781003180241 (v. 5 ; ebook) | ISBN 9781003180289 (v. 6 ; ebook)
Subjects: LCSH: Biotechnology—Nigeria.
Classification: LCC TP248.195.N6 M95 2022 (print) | LCC TP248.195.N6 (ebook) | DDC 660.609669—dc23
LC record available at https://lccn.loc.gov/2021047039
LC ebook record available at https://lccn.loc.gov/2021047040

ISBN: 978-0-367-76705-1 (hbk)
ISBN: 978-1-032-01841-6 (pbk)
ISBN: 978-1-003-18028-9 (ebk)

DOI: 10.1201/9781003180289

Typeset in Times
by codeMantra

Dedication

This book is dedicated to Professor Vincent Ado Tenebe, FBSN, Chairman of the Board of Trustees (BoT) of the Biotechnology Society of Nigeria (BSN) for his passion for Biotechnology and commitment to the cause of the Society.

Contents

Foreword

Biotechnology Society of Nigeria (BSN), the foremost professional society of Biotechnologists in Nigeria, has over the past 40 years been in the forefront of knowledge creation and transfer in all the areas of biotechnology. This has found expression in a number of major high-quality textbooks and documents published by the Society, particularly in the last few years. These texts cover all the fundamental and modern areas of biotechnology and have been of great value to biotechnologists at undergraduate, postgraduate and professional levels the world over.

The current volume on *Multidisciplinary Applications and Advances of Biotechnology – VI: Bioenergy and Environmental Biotechnology for Sustainable Development* is coming at a time of worldwide and significant turmoil in energy use in the face of global warming, climate change and energy-related pollution. In this volume of eight chapters, authorities of global repute have put together information at the cutting edge of the subject. In Chapter 1, Bio-resources Utilization in Fostering a Low-Carbon Renewable Energy–Based Economy, carbon-neutral and renewable energy sources are discussed in the light of the challenges of fossil fuel and climate change. Bioenhancement formulations for environmental clean-up form the subject of Chapter 2. This is particularly important for global fossil fuel producers such as Nigeria where environmental degradation and challenges of pollution have combined to negatively impact food security and the cost of energy production, while exacerbating community restiveness in oil producing areas. Biomethanation is the subject of Chapter 3. The authors discuss methane generation in anaerobic digesters as a way to release energy trapped in various organic wastes. Practical considerations for the successful operation of anaerobic digesters under different environmental and biochemical conditions are discussed. Chapter 4 looks at strategies for successful in-situ and ex-situ remediation of industrially polluted environment, by employing biological and combined biological and physical or biological and chemical processes. In Chapter 5, experts explore the evolution and current status of environmental reclamation in Nigeria. This is particularly topical in the light of considerable pollution that has accompanied oil exploration and production activities in very sensitive ecological regions of Nigeria. The chapter highlights reclamation methods applied within the Nigerian environmental space over time and suggests improvements. Microbial remediation of plastic pollution is the subject of Chapter 6. The chapter explores prospects and challenges in deploying the physiological versatility available in the microbial world for managing plastic pollution. This is particularly topical in the face of the growing place of petroleum-derived plastics in the modern world and the challenge of plastics-associated pollution. In Chapter 7, novel approaches to nanoparticle synthesis are discussed. The emphasis is on environmentally friendly synthesis options (green synthesis) employing renewable (plant) materials. Chapter 8 on the subject of sustainable wastewater treatment explores the use of microalgae in win-win methods for wastewater treatment with the production and recovery of various algal-based, value-added products.

The authors are leading authorities in the field. This text will be an indispensable resource for biotechnology professionals as well as undergraduate and postgraduate students interested in environmental biotechnology.

Jerry O. Ugwuanyi, PhD (Strathclyde, UK)
Professor of Microbiology
Dean, Faculty of Biological Sciences
University of Nigeria (2016–2018)

Preface

In recognition of biotechnology's pivotal role as a burgeoning technology in fostering a knowledge-driven bio-based economy, the Biotechnology Society of Nigeria (BSN) has – in the about 40 years of its founding – made enormous contributions towards expanding the frontiers of biotechnology in Nigeria for sustainable livelihoods. This book is one of the multi-volume series on the Multidisciplinary Applications and Advances in Biotechnology: Contributions from the Biotechnology Society of Nigeria (BSN), with each volume covering specific contemporary topics targeted at providing scientists in the academia, industry, research institutes/agencies, policy makers, etc. with recent advances in the field to bridge the knowledge gap in the crucial areas of biotechnology.

This book is designed to document advances in biotechnology in Nigeria over the past decades and the emerging biotech sector as a key driver of the bioeconomy. It is a timely knowledge product detailing key issues on recent developments in the field of biotechnology particularly in Nigeria. It is written for use by various readers including postgraduate students, professionals in the area of research, industry players, policy makers, educationists, science advocacy groups, communication experts, etc. Additionally, the book engenders a deeper understanding of underlying developments in biotechnology, its historical perspectives and context in Nigeria, processes and science-policy nexus that have placed Nigeria as a leader in biotechnology in Africa.

This volume covers 2 of the 16 Working Groups of the Society, namely Waste Recycling, Biofuels and Environmental Remediation (WG09) and Water Research (WG16). It contains important topics on environmental remediation; biofuels and value-added microbial products for environmental clean-up; and water and waste water recycling which are spread in nine chapters that bring to light the applications of available knowledge and deliverables in the thematic areas covered by the array of seasoned contributing authors. The chapters in the Waste Recycling, Biofuels and Environmental Remediation section cover a wide range of topics including the utilization of bio-resources in fostering a low-carbon renewable energy–based economy; biomethanization in anaerobic systems: feedstock and process enhancement; bioactivity potentials of eco-friendly synthesized metallic nanoparticles; use of enhanced bioprocesses for the reclamation of impacted environment; potential of some indigenous bacterial isolates in the development of bioenhancement formulations for bioremediation of hydrocarbon impacted environments; environmental reclamation in Nigeria: past, present and future; and the role of microorganisms in curbing plastic and petrochemical menace. In the Water Research section, the book covers sustainable wastewater treatment using microalgae technology and new insights into chlorine disinfectant usage in water quality: a need for review of existing guidelines.

Each chapter addresses the relevant points in the subject areas covered and discusses each of the thematic areas concisely. The chapters are co-authored by a team of different scientific experts in the field which guarantees in-depth scholarly output in a high level of interdisciplinary collaborative manner. We have leveraged the pool of expertise available in the two BSN Working Groups to ensure a higher level of sustainability on the deliverables expected from these thematic research areas.

This book is a useful reference material, a comprehensive description of the recent advances in the field and a valuable guide to the subject matter's original literature. It is specifically targeted at scientists and researchers working in the different areas of biotechnology including bioenergy for a circular bioeconomy, environmental remediation, formulation of sustainable and environmentally safe bioenhancement processes from indigenous bio-resources, microalgae technologies and water treatment, educationists, green environment advocacy groups, etc. This book will be valuable to professionals in universities, research institutions, industry or government agencies and policy makers/science advocacy groups.

Overall, this volume is an excellent piece of scientific contributions that targets a wide range of audiences including policy makers, private sector, researchers, civil society, farmers and fishing communities, government and environmentalists, which could help in fostering an innovative pathway for future sustainable environment that could ensure availability of safe water, resilient and healthy environments to meet the needs of an increasing global population. It should also be of relevance to industrialists with interest in developing Africa's indigenous bio-resources and biotech start-ups, environmental biotechnology sector, professionals involved in the transformation of waste to wealth, educators and researchers, and educational and research institutions. Readers will find this book a valuable resource containing useful information in the thematic areas contributed by a seasoned and diverse array of scholars within this professional society.

The carefully selected editorial team and the authors of individual chapters drawn from several key institutions were chosen for their recognized international expertise and contributions to the various subject areas. Their willingness to make these invaluable contributions in their respective fields of expertise reflecting the different Working Groups and in a collaborative way to foster research networking is gratefully acknowledged. The editors wish to thank the different manuscripts' reviewers for their unceasing dedication and significant input. We thank Prof. Malachy O. Akoroda of the Department of Agronomy, University of Ibadan, who shared the initial idea that inspired this book initiative with me just after the conclusion of the 32nd Annual Conference of the Society held at the International Institute of Tropical Agriculture (IITA) Ibadan, Nigeria. Furthermore, we highly appreciate the foresight, diligent support and understanding of the publisher, Taylor and Francis Group (USA), for publishing *Multidisciplinary Applications and Advances in Biotechnology – VI: Bioenergy and Environmental Biotechnology for Sustainable Development* with their international hallmark of excellence. We return special thanks to Dr. Marc Gutierrez and Mr. Nick Mould, whose tremendous and painstaking efforts, constant suggestions and guidance at all stages led to the successful publication.

Prof. Benjamin Ewa Ubi
President, Biotechnology Society of Nigeria (BSN)
& Chief Editor
24th March, 2021

Editors

Akinola Rasheed Popoola, Ph.D., is a Professor of Plant Pathology and the Director, Biotechnology Centre, Federal University of Agriculture, Abeokuta, Nigeria. With B.Sc./M.Sc. in Biochemistry and Ph.D. in Plant Pathology, Prof. Popoola had applied the knowledge of molecular biology in addressing plant disease issues. His past work had explored molecular markers (RFLP, SSR, CAPS) to detect resistant genes to wilt and other diseases in tomato. He had collaborated with scientists within and outside Nigeria to improve tomato genotypes with respect to disease resistance. Prof. Popoola was the Principal Investigator in a number of funded projects and the past Editor-in-Chief of the *Nigerian Journal of Biotechnology*. He has over 70 articles in journals and conference proceedings.

Emeka Godfrey Nwoba, Ph.D., is currently a research Fellow at the Algae Research & Development Centre, Murdoch University, Western Australia, Australia. His research areas include industrial & environmental biotechnology, microalgae biotechnology, applied phycology/algology, energy-efficient photobioreactors (PBRs) and biostatistics. He has over 6 years of research and development experience in algae (microalgae and macroalgae) biotechnology including algae-based wastewater treatment, bio-based products from algae, and design, optimization and scale up of PBRs. His academic training was at Ebonyi State University, Abakaliki, Nigeria, where he received his B.Sc. and M.Sc. degrees, in Biochemistry/Industrial Biotechnology. Dr. Nwoba obtained his PhD in Biotechnology/Energy from Murdoch University, Australia. He has published several peer-reviewed research and review articles in Q1 Journals and book chapters. His current research focuses on partnering engineering, material science and biology to develop high efficiency/next-generation algal PBRs (i.e. autonomous/standalone PBRs) with the capability to co-produce chemical (algal biomass) and electrical energy towards freshwater-based cooling and grid-electricity independent operation. He is a member of Australasian Society of Phycology and Aquatic Botany, International Society for Applied Phycology, Phycological Society of Nigeria and Biotechnology Society of Nigeria.

James Chukwuma Ogbonna, Ph.D., is a Professor of Industrial Microbiology and Biotechnology. He obtained a B.Sc. (First-Class Hons in Botany) from University of Jos, Nigeria (1982); M.Eng. (Fermentation Technology) from Yamanashi University, Japan (1988); and Ph.D. (Applied Biochemistry) from University of Tsukuba, Japan (1991). He is currently the Director of National Biotechnology Development Agency, South East Zonal Biotechnology Centre, University of Nigeria, Nsukka, and the immediate past Deputy Vice Chancellor (Academic), University of Nigeria Nsukka. He was the National President, Biotechnology Society of Nigeria (2010–2014) and the current President, Foundation for African Development through International Biotechnology. His current research interests include photobioreactor design and optimization, development of microbial cell culture systems and applications, and bioenergy production. He has published 3 books, co-edited 3 books, published 13 book chapters and 111 peer-reviewed journals articles, and presented 9 conference proceedings and over 120 papers in national and international conferences in over 11 countries. His current H-index is 34 with over 3,760 citations in Google Scholars. Prof. Ogbonna is a Fellow of the Biotechnology Society of Nigeria, Fellow of the Foundation for African Development through International Biotechnology, and Fellow of the Institute of Corporative Administration.

Charles Oluwaseun Adetunji, Ph.D., is an Associate Professor of Microbiology and Biotechnology at the Microbiology Department, Faculty of Science, Edo State University UzairueEDSU, Edo State, Nigeria, where he utilized the application of biological techniques and microbial bioprocesses for the actualization of sustainable development goals and Agrarian revolution, through quality teaching, research and community service. He is currently the Acting Director of Intellectual Property and Technology Transfer, Sub-Dean for Faculty of Science, and the Head of department of Microbiology at EDSU. He is a Visiting Professor and the Executive Director for the Center of Biotechnology, Precious Cornerstone University, Ibadan. He has won several scientific awards and grants from renowned academic bodies like Council of Scientific and Industrial Research (CSIR), India; Department of Biotechnology (DBT), India; the World Academy of Science (TWAS), Italy; Netherlands Fellowship Programme (NPF), the Netherlands; the Agency for International Development Cooperation; Israel; and Royal Academy of Engineering, UK, among many others. He has filed several scientific patents on Bioherbicides, Biopesticides, Nanobiosurfactants, Nanobiopesticdes and many more. He has published over 300 scientific articles in refereed journals and conference proceedings both international and local journals. He was ranked among the top 500 prolific authors in Nigeria between 2019 till date by SciVal/SCOPUS. He is currently editing several biotechnology text books with Elsevier. He was recently appointed as the President and Chairman Governing Council of the Nigerian Bioinformatics and Genomics Network Society. He is presently a series editor with Taylor and Francis, USA editing several textbooks on Agricultural Biotechnology, Nanotechnology, Pharmafoods, and Environmental Sciences. The breadth of his scholarly contributions to research is evident from his contributions which cover several topics relating to Food Security, Agriculture and Environmental Sustainability. His research interest includes microbiology, biotechnology, bioresources and nanotechnology. He is an editorial board member of many international journals and serves as a reviewer to many double-blind peer-review journals like Elsevier, Springer, Taylor and Francis, Wiley, PLOS One, Frontiers, Nature, American Chemistry Society, Bentham Science Publishers, etc. He is a member of many scientific and professional bodies including the American Society for Microbiology, Nigerian Young Academy, Biotechnology Society of Nigeria and Nigerian Society for Microbiology. He has won a lot of international recognition and also acted as a keynote speaker delivering invited talks/position papers at various universities, research institutes and several centres of excellence which span across several continents of the globe. Over the last 15 years, he has built strong working collaborations with reputable research groups in numerous and leading universities across the globe. He is the convener for Recent Advances in Biotechnology, which is an annual international conference where renowned Microbiologists and Biotechnologists come together to share their latest discoveries (https://raibconference.wixsite.com/website). He is the Founder & CEO of BECTIK Biotechnology and Nanotechnology Company. His Biotechnology Company consists of a team of leading academics in their fields and thrives to deliver solutions all around Bio- and Nanotechnology. The company also delivers solutions that are tailored to your business needs and for any scale.

Nwadiuto (Diuto) Esiobu (Ph.D., University of Louvain, Belgium; Post Doc. Massachusetts Institute of Technology, Cambridge, USA) is currently a Professor of Microbiology and Biotechnology at Florida Atlantic University, Boca Raton, FL, USA, where her microbial biotech laboratory integrates environmental genomics and health risk assessment with rapid pathogen detection using synthetic biology. Her research interests in agriculture and health include molecular ecology of plant invasion, soil fertility and the applications of microbiomes in humans and plants. She has served in senior professional leadership roles in academia, industry and policy, including selection panel for the US President's award for excellence in Math and Science (Washington, DC) and more than 11 years on Committees of the Board of the American Society for Microbiology. A Fellow of the Biotechnology Society of Nigeria and a Jefferson Science Fellow, she was selected by the National Academy of Sciences as a senior science advisor to Secretary Clinton's office of Global Food Security (GFS) and the Office of International Health and Biodefense (OES/IHB), US Department of State, Washington, DC, and named world-class Faculty by the City of Fort Lauderdale for her contributions to the scientific enterprise.

Abdulrazak Ibrahim, Ph.D., is an agricultural biotechnologist with experience and skills on plant biotechnology stewardship, stakeholder engagement, foresight analysis, development of transgenic crops, laboratory and field management of GM crops, biosafety analysis, and adoption and discontinuation of events. He was trained in the Brazilian Agricultural Research Cooperation (EMBRAPA Center for Genetic Resources), from where he obtained a doctorate degree in Molecular Biology from Universidade de Brasilia (UnB) and M.Sc. from Universidade Federal do Ceara (UFC), Fortaleza, and acquired best practices in agricultural biotechnology leading to the development and patenting of a transgenic technology for the control of whitefly using RNA interference technology within EMBRAPA, and the development of virus-resistant cowpea, under the Africa-Brazil Innovation Platform, implemented by EMBRAPA and Forum for Agricultural Research in Africa (FARA), with support from the Department for International Development (DFID), Bill & Melinda Gates Foundation (BMGF), International Fund For Agricultural Development (IFAD) and Foundation Arthur Bernardes (FUNARBE).

Benjamin Ewa Ubi, Ph.D., is a Professor of Plant Breeding and Biotechnology. He obtained his B.Sc. Honours (Agric.) Agronomy from the University of Calabar (1990) and M.Sc. Agronomy – Crop Science (Plant Breeding and Tissue Culture) (1994) and Ph.D. Agronomy (Plant Breeding and Biotechnology) (1998) from the University of Ibadan, Nigeria. His Ph.D. research was completed at the then Biotechnology Laboratory of the International Institute of Tropical Agriculture (IITA), Ibadan, where he was a graduate research fellow and became the President of the International Association of Research Scholars and Fellows (IARSAF) in 1996. Among several other research engagements, Prof. Ubi served as Science and Technology Agency (STA) Postdoctoral Fellow under the Japan International Science and Technology Exchange Centre (JISTEC) 1999–2001 at the Plant Breeding Department (Biotechnology Laboratory) of the National Institute for Livestock and Grassland Science (NARO, National Agricultural Research Organization), Nishinasuno, Japan; and as a three-term Visiting Scientist at the NARO National Institute of Fruit Tree Science (NIFTS), Tsukuba, Japan, under the auspices of the Inoue Research Foundation of Japan Fellowship (April 2003–April 2004) and the Japanese Society for the Promotion of Science (JSPS) visiting fellowships (October 2008–October 2009; May 2012–March 2013). Prof. Ubi served as Visiting Professor at the Arid Land Research Centre (The Global Centre of Excellence for Dryland Science) (Molecular Breeding Laboratory), Tottori University, Japan; and Visiting Professor University of Tokyo, Japan. Among several other academic engagements, Prof. Ubi has served as the Director of the Biotechnology Research and Development Centre, Ebonyi State University, Abakaliki (August 2006–March 2008) and the Head of the Department of Biotechnology, Ebonyi State University for several years; Dean, Faculty of Science, Ebonyi State University Abakaliki and Pioneer Editor-in-Chief of *Science Journal* <https://sciencejournal.ng>, a publication of the Faculty of Science, Ebonyi State University, Abakaliki; President, Biotechnology Society of Nigeria (BSN) and Leader of the Crops Genetics, Breeding and Genetics Working Group of the Biotechnology Society of Nigeria (BSN).

Contributors

I. K. Adam
Federal University Lafia
Lafia, Nigeria

Isaac J. Adekunle
University of Benin
Benin City, Nigeria

A. O. Adeleye
Department of Environmental Sciences
Federal University Dutse
Dutse, Nigeria

Geoffrey O. Anoliefo
Environmental Biotechnology and Sustainability Research Group
Department of Plant Biology and Biotechnology, Faculty of Life Sciences
University of Benin
Benin City, Nigeria

Modupe Stella Ayilara
Food Security and Safety Niche Area, Faculty of Natural and Agricultural Science
North-West University
Mmabatho, South Africa
and
Genetic Resources Centre
International Institute of Tropical Agriculture
Ibadan, Nigeria

Mathias A. Chia
Department of Botany
Ahmadu Bello University
Zaria, Nigeria

N. C. Ezebuiro
National Biotechnology Development Agency (NABDA)
Abuja, Nigeria

A. A. Farouq
Department of Microbiology
Usmanu Danfodiyo University
Sokoto, Nigeria

John N. Idenyi
Department of Biotechnology
Ebonyi State University
Abakaliki, Nigeria

Francis Aibuedefe Igiebor
Environmental Biotechnology and Sustainability Research Group
Department of Plant Biology and Biotechnology, Faculty of Life Sciences
University of Benin
Benin, Nigeria
and
Department of Microbiology, Faculty of Natural and Applied Sciences
Wellspring University
Benin City, Nigeria

Beckley Ikhajiagbe
Environmental Biotechnology and Sustainability Research Group
Department of Plant Biology and Biotechnology, Faculty of Life Sciences
Applied Environmental Biosciences and Public Health Research Group
Department of Microbiology, Faculty of Life Sciences
University of Benin
Benin City, Nigeria

I. Körner
Technische Universitaat Hamburg, Harburg
Hamburg, Germany

Saheed Ibrahim Musa
Admiralty University of Nigeria
Delta State, Nigeria

M. E. Nkereuwem
Department of Soil Science
Federal University Dutse
Dutse, Nigeria

K. E. Nwagu
Department of Biotechnology
Alex Ekwueme Federal University Ndufu Alike
Abakaliki, Nigeria

Emeka G. Nwoba
Algae R&D Centre
Murdoch University
Murdoch, Australia

Olu Odeyemi
Department of Microbiology
Obafemi Awolowo University
Ile-Ife, Nigeria

Christiana N. Ogbonna
Department of Plant Science and Biotechnology
University of Nigeria
Nsukka, Nigeria

James C. Ogbonna
Department of Microbiology
University of Nigeria
Nsukka, Nigeria

Mike O. Ojemaye
SAMRC Microbial Water Quality Monitoring Centre
University of Fort Hare
Alice, South Africa

Anthony I. Okoh
Department of Environmental Health Sciences
College of Health Sciences
University of Sharjah
Sharjah, United Arab Emirates
and
SAMRC Microbial Water Quality Monitoring Centre
University of Fort Hare
Alice, South Africa

Oluwaseyi Samuel Olanrewaju
Food Security and Safety Niche Area, Faculty of Natural and Agricultural Science
North-West University
Mmabatho, South Africa
and
Genetic Resources Centre
International Institute of Tropical Agriculture
Ibadan, Nigeria

Ogochukwu F. Olise
Environmental Biotechnology and Sustainability Research Group
Department of Plant Biology and Biotechnology, Faculty of Life Sciences
University of Benin
Benin, Nigeria
and
Department of Biological Sciences
Anchor University
Lagos, Nigeria

V. O. Onokebhagbe
Department of Soil Science
Federal University Dutse
Dutse, Nigeria

C. O. Onyia
Department of Biological Sciences
Godfrey Okoye University
Enugu, Nigeria

Mojisola Owoseni
Department of Microbiology
Federal University of Lafia
Lafia, Nigeria

Rebecca Richard
Department of Biological Sciences
Federal University Gashua
Gashua, Nigeria

Mobolaji Adenike Titilawo
Department of Microbiology
Osun State University
Osogbo, Nigeria

A. F. Umar
Department of Microbiology
Abubakar Tafawa Balewa University
Bauchi, Nigeria

M. B. Yerima
Department of Microbiology and Biotechnology
Federal University Dutse
Dutse, Nigeria

1 Bio-resources Utilization in Fostering a Low-Carbon Renewable Energy–Based Economy

Modupe Stella Ayilara and
Oluwaseyi Samuel Olanrewaju
North-West University
International Institute of Tropical Agriculture

Olu Odeyemi
Obafemi Awolowo University

Mobolaji Adenike Titilawo
Osun State University

CONTENTS

DOI: 10.1201/9781003180289-1

1.1 INTRODUCTION

Energy is an essential factor for social, rural, economic and security development. It is crucial in the day-to-day activities, and its role cannot be underestimated in the industrial, agricultural, health and education sectors, as well as in the home. In homes, electricity aids conducive temperature, the use of home appliances, cooking and lighting. In the health agencies, it is used to run the machinery, medical equipment and cooling devices, especially in preserving some vaccines and lighting (Franco et al., 2017). In the education parastatal, it is used to operate teaching aids, computers and computer accessories. In agriculture, it is needed to operate machinery used to irrigate fields, process seeds and in fertilizer production. Energy is also essential in transportation, communication, mining, commerce and manufacturing (Østergaard et al., 2020). Energy sourced from fossil fuel has been the major energy source over the years (Zou et al., 2016). This source of energy poses a pollution threat to the environment, and it is also not renewable. This has led to the incorporation of a safer and sustainable energy source. The seventh Sustainable Development Goal (SDG) focuses on affordable, sustainable and clean energy (Acheampong et al., 2017, Gunawan et al., 2020). Sustainability is a crucial factor in determining an alternative to energy production, as the world population is ever-increasing, consequently increasing the primary energy demand (Acheampong et al., 2017). This chapter discusses the different sources of energy and their advantages and disadvantages.

1.2 SOURCES OF ENERGY

Based on the capacity for renewability, energy sources can be classified as renewable and nonrenewable. While oil, coal and natural gas are examples of nonrenewable energy, wind, solar and hydro-based energies are renewable.

1.2.1 Nonrenewable Energy

Generally, nonrenewable energy sources generate a higher amount of power on request; however, they get exhausted over time, are not replenished over a long period of years and generate carbon dioxide and other greenhouse gases (Zhang et al., 2017). About 80% of the world's energy comes from fossil fuels (Harjanne and Korhonen, 2019). Examples of nonrenewable energy are discussed below.

1.2.1.1 Coal

Coal occurs naturally from decaying plants, animals and microbes (Iqbal et al., 2017). These processes take a long time to occur. Electricity is produced from coal by combusting it to produce steam which turns water to steam, and the steam runs the turbines that enhance the generator to form electricity. Coal is the most significant source of fossil fuel, and it's a primary source of energy. It is mostly used by a lot of processes in the manufacturing industry. During coal combustion, carbon is released, and the released carbon reacts with atmospheric oxygen, leading to carbon dioxide production. Carbon dioxide has been highly implicated in the pollution of the atmosphere, leading to ozone layer depletion, greenhouse effects and climate change.

Coal is expected to last for 250 years; therefore, it is not sustainable over a long period making it essential to replace with other energy sources (Odeyemi, 2014).

1.2.1.2 Petroleum Oil

Petroleum oil is produced as a result of the compaction of dead plants and animals under high temperatures and pressure. Oil is used to produce electricity and to power vehicles, generators and other machineries. Compared to other sources of energy, oil is readily available and produces a high amount of energy. The combustion of oil and the emission from vehicles, generators, machineries, aircraft and ships lead to air pollution due to greenhouse gas release (e.g. CO_2, CH_4 and N_2O). Aside from the release of greenhouse gases, environmental pollution could arise through oil spillage. Oil spillage is the accidental or deliberate release of hydrocarbons into the environment, and it could happen during the transfer of crude oil into vessels and tankers, crude oil exploration, oil pipeline vandalization and leakage, and the illegal tampering with oil-well heads (Ndimele et al., 2018). It could have grievous consequences on the environment, and if not well managed, it could lead to fire outbreaks. It can also lead to the death of aquatic lives ranging from plants to animals, fishes, microbes and other aquatic inhabitants or render the soil unuseful for agricultural purposes. Oil is expected to last for about 200 years and the exploration started in 1860, and therefore, by 2060, it will likely be exhausted (Odeyemi, 2014).

Natural gas is a cheap source of energy formed from the decomposed organic wastes acted on by microorganisms over a long period. It differs from crude oil by its molecular weight. It is a complex combination of many gases such as butane, ethane, methane and propane. Natural gas can be used in cooking, heating and electricity production. The greenhouse gases produced from natural gas are much lower than those produced from oil and coal (Annual Energy Outlook, 2015; Koh et al., 2016). The combustion of natural gases leads to the release of small quantities of mercury, particles and sulfur. Methane has the highest proportion of all the gases tagged natural gas. Methane is odorless, tasteless and colorless but undergoes complete combustion and leaves behind no ashes. It is highly flammable and hence is mixed with some aromatic chemicals before it is supplied for domestic use. This is done so that if there is any leakage, it can be easily identified to avert fire disasters.

1.2.1.3 Nuclear Energy

Nuclear energy is produced from nuclear fission, fusion or decay reactions to generate heat which is subsequently used to produce steam from water, turn a turbine and generate electricity (Petrescu et al., 2016). Of all these processes involved, nuclear fission is the most utilized process where uranium and plutonium are utilized (von Hippel et al., 2019). Energy from nuclear sources can be used for household, industrial and agricultural activities and to power electric vehicles. Out of all the nonrenewable sources, nuclear energy is still the most preferable because it does not lead to the release of greenhouse gases. Unfortunately, their drawback lies in the fact that they can get depleted, and they produce radioactive elements that could be dangerous to humans and the environment. The carbon released from nonrenewable energy ranges from 22% in coal, 33% in petroleum, to 22% in natural gas (Ramseur, 2017; EPA, 2017).

1.2.1.4 Advantages of Nonrenewable Energy

Nonrenewable energy is easily accessible, and the amount of energy they generate is huge. Nonrenewable energy is diverse in end products, and the diversity is seen in the refinery process leading to the production of different products such as plastic and soap. Fossil fuels recombined with some chemicals are used to produce many day-to-day household items such as plastics, floor tiles, detergents, synthetic fibers and fertilizers (Dickson, 2016). To a large extent, fossil fuel is still the major source of energy, especially in the developing world. It holds the global economy and is easier to transport than their renewable counterparts; it is also used in clean energy production.

1.2.1.5 Disadvantages of Nonrenewable Energy

Nonrenewable energy leads to the release of greenhouse gases which leads to global warming and diseases in human. It takes time to search, mine and drill fossil fuel. The energy source is capable of vanishing; therefore, it is not sustainable. Another disadvantage is the environmental pollution caused by fossil fuel. In areas where oil is being drilled, oil spillage occurs. This, in turn, affects the livelihood of the immediate environment by disrupting farming activities and posing an unnecessary pressure to increase prices of oil (owing to higher demands, no solid alternatives and lower availability). Mining nonrenewable energy sources is a tedious process and puts workers through many health and occupational hazards.

1.2.2 Renewable Energy

These are energy derived from the wind, sunlight, water, plant, animals and geothermal sources (Mohtasham, 2015). They do not directly release carbon dioxide into the atmosphere, thereby reducing environmental pollution. They are sustainable sources of energy since their origins are infinite; for instance, with the availability of sunlight every day, solar energy is reliable. However, power generation might still be limited when the sky is cloudy. As of 2017, about one-quarter of all electricity generated worldwide is from renewable sources (Gielen et al., 2019). Several countries are embracing the idea of renewable energy. For instance, countries like Vietnam, Bangladesh, Colombia, Ghana, Barbados, Ethiopia, Cambodia, Mongolia and Hawaii proposed that 100% of their energy will be from renewable sources by 2045 (Hansen et al., 2019).

Renewable energy could exist in the solid, liquid or gaseous form. The solid state could be in the form of wood, charcoal, pellets and chips (Perea-Moreno et al., 2016). The liquid form includes biodiesel and bioethanol, while the gaseous form (biogas) can be obtained through gasification (semi-oxidation of biomass at high temperature) or anaerobic fermentation to produce gases such as nitrogen, methane and hydrogen (Weiland, 2010; Perea-Moreno et al., 2019). The positive economic and environmental effects of renewable energy have been reported both in a long and short run (Haseeb et al., 2019). Bulavskaya and Reynès (2018) reported that renewable energy would add about 1% to the gross domestic product and create about 50,000 new jobs in Dutch by 2030. The utilization of renewable energy will further help reduce greenhouse gas emissions, consequently improving human health and economy growth (Khan et al., 2020). Charfeddine and Kahia (2019) researched the environmental and economic effects of renewable energy in the East and North Africa from 1980 to 2015, and they found out that both facts were slightly affected, indicating a weak development in the renewable energy sectors. Sources of renewable energy are discussed in detail below.

1.2.2.1 Solar Energy

This energy is generated by trapping and converting radiation from the sun into hot water, heat or electricity. Solar energy is one of the freely available sources of energy. It is available in abundance and generated by trapping and converting radiation from the sun into hot water, heat or electricity.

This involves the harvesting and use of light or heat energy through active and passive technologies (Besarati et al., 2013). Passive technology involves the direct trapping of solar (thermal or light) energy for energy generation; such energy can be used for heating during winter (Sun and Wang, 2016). Active solar systems trap solar radiations and convert it to heat and can be used to power equipments such as fans and pumps (Kabir et al., 2018). There are two types of active solar energy technology, namely, solar thermal technology and photovoltaic technology (Herrando and Markides, 2016). The active solar energy technology is used to convert sunlight energy into thermal energy, and the generated thermal energy is thus used for cooking, drying, cooling, etc. (Seddegh et al., 2015). The photovoltaic technology, which is the more preferred option, involves semiconductors to convert sunlight directly into electricity (Mohanty et al., 2016).

The earth receives about 342 W m^2 of energy from the sun, with about 30% reflected back, leaving behind 70% (239 W m^2) for harvesting and conversion (Hart, 2015). Annually about 60–250 W m^2 of solar irradiance is effective throughout the world (Luqman et al., 2015). Africa is considered the sunniest place globally, receiving a concentrated solar power and photovoltaic energy of about 470 and 660 PW, respectively (Iram et al., 2020). Other locations aside from Africa, e.g. Australia, Southwestern United States, the desert plains of India, Central and South America, Middle East, Pakistan and Northern and Southern Africa, only generate 125 GWh from 1 km^2 area of land (Adaramola, 2014). Australia accounts for the most efficient solar energy worldwide, owing to the high solar radiation per square meter, which is the highest in the world (Kabir et al., 2018). The country receives radiation of 4 KW h/m^2 daily (Baniyounes et al., 2011).

The adaptation of solar energy has helped a lot to reduce the emission of 696, 544 metric tons of CO_2 emission (Arif, 2013). Solar energy is not associated with the release of greenhouse gases and particles (carbon black, particulate matter, soot or metals) (Kabir et al., 2018). The drawbacks of this source of energy are that it is subject to the weather condition, also the initial cost for installation is high, and the system's efficiency can be reduced by water intrusion, cracks in the PV module and algal growth (Kabir et al., 2018).

1.2.2.2 Wind Energy

Turbines are used to trap wind flow and turn the wind energy into electricity. Over 420 GW of energy was produced in 2016 using wind energy, and this value is expected to rise to 1,000 in 2030 (GWEC, 2011; Marugán et al., 2018). Energy produced from wind accounts for about 28% of the world's energy demand, and even with this, very few countries (e.g. China) have indicated an interest in it (Sitharthan et al., 2018).

The generation of wind power worldwide has increased tremendously in the past few decades, with about 30% per year (Leung and Yang, 2012). Countries like India, China, Germany, Spain and the USA take the lead when it comes to the ranking of wind energy utilization worldwide (Sahu, 2018). The Global Wind Energy Council predicted a 17% increase in global wind energy generated worldwide (Lu and McElroy, 2017). Archer and Jacobson (2005) proposed that 20-folds of the total wind energy generated globally could be up to 123 PWh of electricity annually, comparable to seven times the total current global electricity demand. This makes wind energy a promising source of sustainable energy.

There are different types of turbines that convert wind energy into electricity such as horizontal-axis turbines, vertical-axis turbines and ducted wind turbines. This energy source is not associated with pollution because it does not produce carbon dioxide; however, it is subject to uncertainties due to weather conditions and climate change. Other drawbacks of wind energy include noise and visual pollution, though this impact seems negligible when compared to the effects of fossil fuels. It is therefore vital to carry out more research to rectify these drawbacks to enhance the sustainability of wind energy (Leung and Yang, 2012).

1.2.2.3 Hydro Energy

Electricity is produced from water released from a dam which will drive a turbine to produce electricity (Mohtasham, 2015). Water is pushed from an elevated position of the dam through a pipe, and the water moves from the reservoir to the turbine. The pressure arising from the flow of water is driven to the blades of the turbine, and thus rotates the electric generator's rotor. The coils in the wire of the rotor rolls pass the generator's static coil, leading to electricity generation. This source of energy is still more dependable than solar energy due to the availability of water. It is cheaper, is not associated with the generation of carbon dioxide and other greenhouse gases, and thus does not directly cause any environmental pollution (Igliński, 2019). It produces 19% of the electrical power used globally (Igliński, 2019). A number of countries have ventured into hydroelectric energy production. For instance, Poland and Colombia have been using hydroelectric power for more than 100 years, with the former generating 784.44 MW currently and the latter is proposed to generate 125 GW (Gómez-Navarro and Ribó-Pérez, 2018). France has been reported to be one of the biggest hydroelectric power producers (Larinier, 2008; Pang et al., 2015). Close to half of the energy used in Sweden comes from hydropower, making it the most important source of energy (Manzano-Agugliaro et al., 2017). In Norway, the energy production of about 300 GWh has been reported (Manzano-Agugliaro et al., 2017). The major disadvantages of hydropower are the changes in the water levels, water currents and movement pathways, and the changes in the paths and directions of the inhabitants of the water bodies (Singh and Singal, 2017).

1.2.2.4 Geothermal Energy

Geothermal energy is produced from the heat generated deep within the earth's crust due to radioactive decay and the process of earth formation (Østergaard et al., 2020). Steam generated from the pumping of hot water deep within the earthcrust which is rising to the top of the earth is used to operate turbines that traps energy.

There is about 3.6×10^{14} GWh of energy stored at the upper 10 km of the earth's crust (Lund, 2007). According to the 2012 global energy consumption rate of around 1.7×10^{8} GWh, the geothermal energy source might supply the total energy demand globally for about 2.17 million years (BP, 2016; Lu, 2018). Geothermal energy source is capable of serving as a sustainable source of energy. The USA, the Philippines, Indonesia, Mexico, New Zealand, Italy, Iceland, Kenya, Japan and Turkey are the top 10 highest geothermal energy generating countries (Hou et al., 2018). However, some other countries (Malaysia, Croatia and Iran) are gradually adopting the idea of geothermal energy.

The advantages of this form of energy are that it is not affected by weather conditions, it can generate energy at any time and it can help to prevent disasters by trapping the energy that could cause disasters such as geysers and volcanic eruptions. It is quite expensive to set up and is prone to earthquakes over a long time.

1.2.2.5 Biomass Energy

Biomass energy is produced from living things, specifically plants and organisms. This can be produced by thermal combustion, direct conversion or conversion into biofuel. Globally, roughly 146 billion metric tons of biomass is produced annually (IPCC, 2011). Around 5% of the primary energy in industrialized countries comes from biomass, while 35% of the primary energy used in the developing countries comes from biomass, making a total of 14% world reliant on biomass (IPCC, 2011; Toklu, 2017).

The energy could be produced through gasification, direct-firing, pyrolysis, torrefaction (thermal combustion of biomass to a temperature above 200°C) and co-firing (Goerndt et al., 2013). Gasification is done to produce syngas and slag by heating municipal solid waste to more than a temperature of 700°C using oxygen in a controlled manner. Slag is a molten liquid that has a resemblance to gas. It is used as a raw material to produce asphalt, cement and shingles. Syngas is made up of hydrogen and carbon monoxide without pollutants (e.g. sulfur) (Kang et al., 2020). It is used as a raw material to produce chemicals, fertilizers and biofuels or burned to produce electricity and heat. The biomass produced gets dried beyond spoilage, becomes black and is compressed to form briquettes. Biomass can be combusted with fossil fuel or briquette (co-firing) to produce energy. The co-firing helps to reduce the quantity of greenhouse gases emission. Pyrolysis is the process of combusting organic wastes in the absence of oxygen, thereby preventing carbon release. The process could generate heat above 200°C, and the end products include solid (biochar), liquid (pyrolysis oil) and gaseous (syngas) products which are used to produce energy. Biofuel can also be categorized as biomass energy, since they come from organic sources (living things). Biofuels are fuel energy obtained as a result of biological carbon fixation or the decomposition of previously living things (Rodionova et al., 2017). These could be photosynthetic organisms (vascular terrestrial plants and macro- and microalgae).

Biofuels include biogas, syngas, bioalcohols, green diesel, biodiesels, algal fuel, bioethers and solid biofuels. Biogas is composed majorly of methane, and it is produced from anaerobic decomposition of organic materials with the aid of anaerobic bacteria. Biogas is used to cook, generate electricity and power vehicles (Odeyemi, 2014). Scarlat et al. (2018) reported that biogas production in the European Union (EU) countries was about 654 PJ (18 billion M^3 methane) in 2015 which is about half of the biogas produced globally. Msimanga and Sebitosi (2014) reported China, Italy, India, the USA and Germany as the top five biogas-producing countries. They reported that out of the five countries, two are EU countries (Italy and Germany), and this indicates tremendous progress in the region which African and other developing nations should replicate.

Biodiesel is produced from fats and oil in a reaction termed transesterification reaction. It can be made from substrates such as sunflower, vegetable oils, hemp, soy, mustard, animal fats, palm oil, jatropha and algae (Gaurav et al., 2013). It can be used to power engines alone or when mixed with mineral diesel. It is always labeled BX when mixed with mineral diesel; the X refers to the percentage of mineral diesel in the mixture, e.g. B10 denotes 10% of mineral diesel in biodiesel (Romano and Sorichetti, 2010). The global production of biofuels is about 35 billion liters (EU Commission, 2006; Khan et al., 2017). Majority of this comes from the three top biodiesel-producing countries, namely, EU, Brazil and the USA (Balat, 2007; Khan et al., 2017).

Bioalcohols are produced through the actions of microbes and enzymes on substrates containing carbohydrates (cellulose, sugars and starches), thereby converting them to carbon dioxide and alcohol (Gordon et al., 2018). Biobutanol, bioethanols and biopropanols can be produced through this process. Bioethanol is produced using sugars from starchy crops (corn, potatoes or other cereals) or from sugar cane or beets (Cardona et al., 2010; Dodić et al., 2009). It can be mixed with gasoline to prevent the emission of greenhouse gases. About 80% of the total global bioalcohol production comes from Brazil and the USA, with an annual output of over 7,000 million gallons and 15,000 million gallons (over half of the whole world production), respectively (Gavahian et al., 2019).

Green diesel is produced by hydrocracking natural oils from plants and animals at high temperatures and pressures using a catalyst (Taromi and Kaliaguine, 2018). In the production process, big molecules are broken down into small chains and retain a green color because due to the yellow and blue solvent which forms the green colour. The green diesel market is growing globally; it increased from 330×10^6 gallons in the year 2011 to about 2.1×10^9 gallons in 2017 (Douvartzides et al., 2019). The Netherlands, the USA, Finland, Singapore and Italy are the highest green diesel producers in the world (Douvartzides et al., 2019).

Microbial fuel cell (MFC) is a device that turns chemical energy from organic substances directly into electrical energy with the aid of microbes. The MFC works by transferring protons by proton exchange and passing electrons from the anode to the cathode, leading to the production of electric energy (Priya and Setty, 2019). The two electrodes are connected by an external circuit that allows the resulting energy utilization. A barrier (proton exchange membrane) exists between the cathodic and anodic chambers, and protons which aid the process are being supplied into the anode chamber. Environmental pollution is prevented by converting the biowaste to MFCs (Gajda et al., 2018). Microbes in the presence of oxygen remove electron while degrading organic materials; the electrons are further transformed to a final receptor, but some other bacteria referred to as "exoelectrogenic bacteria" can transfer electron from organic compound oxidation to other systems which are not within the cell. The exoelectrogenic bacteria are well utilized as catalysts in MFCs (Logan, 2009; Rachinski et al., 2010). In the aerobic compartment containing the cathode, the electron produced reduces the oxygen which combines with the proton to produce water, while in the anode compartment, the electron produced is collected by a graphite cloth (Logan, 2009; Passos et al., 2016).

1.2.2.5.1 Algal Biofuels

Algae produce energy from the photosynthetic process, where they absorb carbon dioxide and release oxygen. A green crude oil is extracted from algae by subjecting it to high pressure and heat; the remnant from the process is further processed to obtain alcohol (Odeyemi, 2014). Algae can yield about 94,000 L of biofuel per acre annually (Ullah et al., 2014). The easy cultivation, the ability to trap atmospheric CO_2 and the ability to grow in wastewater make algae a preferred option than fossil fuels (Adeniyi et al., 2018). The drawback of the algae biofuel production process is that it is expensive to produce.

1.2.2.6 Advantages of Renewable Energy

It is a sustainable energy source since the source cannot be depleted (Owusu and Asumadu-Sarkodie, 2016). Also, the biomass energy of renewable energy (for instance, algal biofuel) is trapped in the organism and can be harvested when needed. Furthermore, they are cost-effective and environment-friendly, as they are not associated with the release of greenhouse gases such as carbon dioxide, nitrogen oxide, and carbon monoxides, and their maintenance requirements are low. Therefore, they can efficiently replace fossil fuels for a better environment. They are clean sources of energy with a minimum amount of secondary waste production. Through mitigation of greenhouse gas emissions, they reduce global warming when used in place of conventional energy sources (Amleh et al., 2019). Gielen et al. (2019) made a convincing point in their study that the global energy transition could directly or indirectly create additional 19 million jobs in the year 2050. Although they claimed that there would be job losses in the fossil sector, they also emphasized that the job opportunities renewable energy production will create will completely offset the losses.

1.2.2.7 Disadvantages of Renewable Energy

Renewable energy can be affected by slight changes in nature. Solar energy is dependent on the intensity of sunlight, while hydro energy and wind energy can be affected by climate change (e.g. a drop in the sea water level and unfavorable wind pattern can affect the efficiency of renewable energy) (Azarpour et al., 2013; Petrescu et al., 2016).

1.3 CONCLUSION AND RECOMMENDATION

The effects of climate change on the environment, humans and animals cannot be overestimated. The consequences include loss of habitats, greenhouse effects and water scarcity. Nonrenewable sources of energy are easily available and abundant in huge quantities, but they can be depleted and are also associated with greenhouse gases release. Renewable energy is produced from natural resources which are renewed by nature. However, some processes in renewable energy production lead to the production of carbon dioxide either directly or indirectly. This includes the combustion of fossil fuel to pump water from storage systems and the production of bio alcohols where carbon dioxide is produced as a by-product. The renewable energy source remains preferred because they are infinite, making them sustainable compared to their alternative, which can get exhausted and also pollute the environment. Some renewable energy production which is associated with very little or no environmental pollution should be elevated and improved to bring about safe and sustainable energy sources.

REFERENCES

Acheampong, M., Ertem, F. C., Kappler, B. and Neubauer, P. (2017). In pursuit of Sustainable Development Goal (SDG) number 7: will biofuels be reliable? *Renewable Sustainable Energy Reviews* 75: 927–937.

Adaramola, M. (2014). *Solar Energy: Application, Economics, and Public Perception*. CRC Press, Technology & Engineering, Boca Raton, FL.

Adeniyi, O. M., Azimov, U. and Burluka, A. (2018). Algae biofuel: current status and future applications. *Renewable Sustainable Energy Reviews* 90: 316–335.

Annual Energy Outlook (2015). *Annual Energy Outlook 2015: With Projections to 2040*. Government Printing Office. US Energy Information Administration, Washington, DC.

Amleh, N., Almuhaini, M. and Khalid, M. (2019) Reliability assessment of standalone wind-based energy storage system using monte carlo simulation. *2019 IEEE Innovative Smart Grid Technologies-Asia (ISGT Asia). IEEE*: 1878–1882.

Archer, C. L. and Jacobson, M. Z. (2005). Evaluation of global wind power. *Journal of Geophysical Research: Atmospheres*: 110: 1–10.

Arif, M. S. (2013). *Residential Solar Panels and Their Impact on the Reduction of Carbon Emissions*. University of California, Berkeley, CA. Retrieved from https://nature.berkeley.edu/classes/es196/projects/final/ArifM_.pdf.

Azarpour, A., Suhaimi, S., Zahedi, G. and Bahadori, A. (2013). A review on the drawbacks of renewable energy as a promising energy source of the future. *Arabian Journal for Science and Engineering* 38: 317–328.

Balat, M. (2007). An overview of biofuels and policies in the European Union. *Journal of Energy Sources, Part B* 2: 167–181.

Baniyounes, A., Liu, G., Rasul, M. G. and Khan, M. M. K. (2011). Review on renewable energy potential in Australian subtropical region (Central and North Queensland). *Advanced Materials Research* 347–353: 3846–3855.

Besarati, S. M., Padilla, R. V., Goswami, D. Y. and Stefanakos, E. (2013). The potential of harnessing solar radiation in Iran: generating solar maps and viability study of PV power plants. *Journal of Renewable Energy* 53: 193–199.

BP. (2016). Statistical review of world energy. *Journal of British Petroleum*, 65: 201.

Bulavskaya, T. and Reynès, F. (2018). Job creation and economic impact of renewable energy in the Netherlands. *Renewable Energy* 119: 528–538.

Cardona, C., Quintero, J. and Paz, I. (2010). Production of bioethanol from sugarcane bagasse: status and perspectives. *Bioresource Technology* 101: 4754–4766.

Charfeddine, L. and Kahia, M. (2019). Impact of renewable energy consumption and financial development on CO_2 emissions and economic growth in the MENA region: a panel vector autoregressive (PVAR) analysis. *Journal of Renewable Energy* 139: 198–213.

Dickson, O. A. (2016) Petrochemicals: upstream to plastic industry using ideas, methods, developments and procedures in developing countries. *39th CSN Annual International Conference, Workshop and Exhibition, 2016 Rivers State University of Science and Technology*, Port Harcourt, Nigeria.: Academic Journals. pp. 1–10.

Dodić, S., Popov, S., Dodić, J., Ranković, J., Zavargo, Z. and Mučibabić, R. J. (2009). Bioethanol production from thick juice as intermediate of sugar beet processing. *Biomass Bioenergy* 33: 822–827.

Douvartzides, S. L., Charisiou, N. D., Papageridis, K. N. and Goula, M. A. (2019). Green diesel: biomass feedstocks, production technologies, catalytic research, fuel properties and performance in compression ignition internal combustion engines. *Energies* 12: 809.

EPA. (2017). *Inventory of US Greenhouse Gas Emissions and Sinks: 1990–2015*. US Environmental Protection Agency, Washington, DC.

EU Commission. (2006). *An EU Strategy for Biofuels*. Communication from the Commission COM 34 final, Brussels, 29 p.

Franco, J., Bogach, V. S., Pérez Arroyo, I. and Lasa, M. (2017). Promoting productive uses of electricity in rural electrification programs, No 28263. The World Bank.

Gajda, I., Greenman, J. and Ieropoulos, I. A. (2018). Recent advancements in real-world microbial fuel cell applications. *Current Opinion in Electrochemistry* 11: 78–83.

Gaurav, K., Srivastava, R. and Singh, R. (2013). Exploring biodiesel: chemistry, biochemistry, and microalgal source. *International Journal of Green Energy* 10: 775–796.

Gavahian, M., Munekata, P. E., Eş, I., Lorenzo, J. M., Khaneghah, A. M. and Barba, F. J. (2019). Emerging techniques in bioethanol production: from distillation to waste valorization. *Green Chemistry* 21: 1171–1185.

Gielen, D., Boshell, F., Saygin, D., Bazilian, M. D., Wagner, N. and Gorini, R. (2019). The role of renewable energy in the global energy transformation. *Energy Strategy Reviews* 24: 38–50.

Goerndt, M. E., Aguilar, F. X. and Skog, K. (2013). Drivers of biomass co-firing in US coal-fired power plants. *Biomass Bioenergy* 58: 158–167.

Gómez-Navarro, T. and Ribó-Pérez, D. (2018). Assessing the obstacles to the participation of renewable energy sources in the electricity market of Colombia. *Journal of Renewable Sustainable Energy Reviews* 90: 131–141.

Gordon, R., Gorodnitsky, I., Promtov, M. A. and Moholkar, V. S. (2018). Processes for increasing bioalcohol yield from biomass. Google Patents.

Gunawan, J., Permatasari, P. and Tilt, C. (2020). Sustainable development goal disclosures: do they support responsible consumption and production? *Journal of Cleaner Production* 246: 118989.

GWEC. (2011). Global Wind Energy Council. Global wind report. Annual market update 2010. Ed. Springer, Berlin, Germany, 2: 12–20.

Hansen, K., Breyer, C. and Lund, H. (2019). Status and perspectives on 100% renewable energy systems. *Energy Policy* 175: 471–480.

Harjanne, A. and Korhonen, J. M. (2019). Abandoning the concept of renewable energy. *Energy Policy* 127: 330–340.

Hart, M. 2015. *Hubris: The Troubling Science, Economics, and Politics of Climate Change*, Lulu.com.

Haseeb, M., Abidin, I. S. Z., Hye, Q. M. A. and Hartani, N. H. (2019). The impact of renewable energy on economic well-being of Malaysia: fresh evidence from auto regressive distributed lag bound testing approach. *International Journal of Energy Economics Policy* 9: 269.

Herrando, M. and Markides, C. N. (2016). Hybrid PV and solar-thermal systems for domestic heat and power provision in the UK: techno-economic considerations. *Applied Energy* 161: 512–532.

Hou, J., Cao, M. and Liu, P. (2018). Development and utilization of geothermal energy in China: current practices and future strategies. *Renewable Energy* 125: 401–412.

Igliński, B. (2019). Hydro energy in Poland: the history, current state, potential, SWOT analysis, environmental aspects. *International Journal of Energy Water Resources* 3: 61–72.

IPCC. (2011). Special reports on renewable energy sources. Summary for policy makers, in: *11th Session of Working Group III of the IPCC*. Abu Dhabi United Arab Emirates, 5–8 May 2011.

Iqbal, J., Lehri, L. A., Shah, S. M. and Nadeem, M. (2017). Combustion analysis of coal using coal water slurry technique. *Engineering Technology Research* 2: 49.

Iram, R., Anser, M. K., Awan, R. U., Ali, A., Abbas, Q. and Chaudhry, I. S. (2020). Prioritization of renewable solar energy to prevent energy insecurity: an integrated role. *The Singapore Economic Review* 66: 1–22.

Kabir, E., Kumar, P., Kumar, S., Adelodun, A. A. and Kim, K.-H. 2018. Solar energy: potential and future prospects. *Journal of Renewable Sustainable Energy Reviews* 82: 894–900.

Kang, J., He, S., Zhou, W., Shen, Z., Li, Y., Chen, M., Zhang, Q. and Wang, Y. (2020). Single-pass transformation of syngas into ethanol with high selectivity by triple tandem catalysis. *Nature Communications* 11: 1–11.

Khan, S., Siddique, R., Sajjad, W., Nabi, G., Hayat, K. M., Duan, P. and Yao, L. (2017). Biodiesel production from algae to overcome the energy crisis. *HAYATI Journal of Biosciences* 24: 163–167.

Khan, S. A. R., Zhang, Y., Kumar, A., Zavadskas, E. and Streimikiene, D. (2020). Measuring the impact of renewable energy, public health expenditure, logistics, and environmental performance on sustainable economic growth. *Sustainable Development* 28: 833–843.

Koh, D.-Y., Kang, H., Lee, J.-W., Park, Y., Kim, S.-J., Lee, J., Lee, J. Y. and Lee, H. (2016). Energy-efficient natural gas hydrate production using gas exchange. *Applied Energy* 162: 114–130.

Larinier, M. (2008). Fish passage experience at small-scale hydro-electric power plants in France. *Hydrobiologia* 609: 97–108.

Leung, D. Y. and Yang, Y. (2012). Wind energy development and its environmental impact: a review. *Renewable Sustainable Energy Reviews* 16: 1031–1039.

Logan, B. E. (2009). Exoelectrogenic bacteria that power microbial fuel cells. *Nature Reviews Microbiology* 7: 375–381.

Lu, S.-M. (2018). A global review of enhanced geothermal system (EGS). *Renewable Sustainable Energy Reviews* 81: 2902–2921.

Lu, X. and Mcelroy, M. B. (2017). *Global Potential for Wind-Generated Electricity*. Wind Energy Engineering. Elsevier, Amsterdam, The Netherlands, pp. 51–73.

Lund, J. W. (2007, June). Characteristics, development and utilization of geothermal resources. *Journal of GHC Bulletin* 28: 1–9.

Luqman, M., Ahmad, S. R., Khan, S., Ahmad, U., Raza, A. and Akmal, F. (2015). Estimation of solar energy potential from rooftop of Punjab government servants cooperative housing society Lahore using GIS. *Smart Grid Renewable Energy* 6: 128.

Manzano-Agugliaro, F., Taher, M., Zapata-Sierra, A., Juaidi, A. and Montoya, F. G. (2017). An overview of research and energy evolution for small hydropower in Europe. *Renewable Sustainable Energy Reviews* 75: 476–489.

Marugán, A. P., Márquez, F. P. G., Perez, J. M. P. and Ruiz-Hernández, D. (2018). A survey of artificial neural network in wind energy systems. *Applied Energy* 228: 1822–1836.

Mohanty, P., Muneer, T., Gago, E. J. and Kotak, Y. (2016). Solar radiation fundamentals and PV system components. *Solar Photovoltaic System Applications*. Springer, Cham, Switzerland, pp. 7–47.

Mohtasham, J. (2015). Renewable energies. *Energy Procedia* 74: 1289–1297.

Msimanga, B. and Sebitosi, A. (2014). South Africa's non-policy driven options for renewable energy development. *Renewable Energy* 69: 420–427.

Ndimele, P. E., Saba, A. O., Ojo, D. O., Ndimele, C. C., Anetekhai, M. A. and Erondu, E. S. (2018). Remediation of crude oil spillage. *The Political Ecology of Oil and Gas Activities in the Nigerian Aquatic Ecosystem*. 1st ed. Elsevier, Amsterdam, The Netherlands, pp. 369–384.

Odeyemi, O. (2014). *Two Centuries of Oil and Gas*. China Universal Academic Services, Beijing, China.

Østergaard, P. A., Duic, N., Noorollahi, Y., Mikulcic, H. and Kalogirou, S. (2020). *Sustainable Development Using Renewable Energy Technology*. Elsevier.

Owusu, P. A. and Asumadu-Sarkodie, S. (2016). A review of renewable energy sources, sustainability issues and climate change mitigation. *Cogent Engineering* 3: 1167990.

Pang, M., Zhang, L., Ulgiati, S. and Wang, C. (2015). Ecological impacts of small hydropower in China: insights from an emergy analysis of a case plant. *Energy Policy* 76: 112–122.

Passos, V. F., Aquino Neto, S., Andrade, A. R. D. and Reginatto, V. (2016). Energy generation in a microbial fuel cell using anaerobic sludge from a wastewater treatment plant. *Journal of Scientia Agricola* 73: 424–428.

Perea-Moreno, A.-J., Juaidi, A. and Manzano-Agugliaro, F. (2016). Solar greenhouse dryer system for wood chips improvement as biofuel. *Journal of Cleaner Production* 135: 1233–1241.

Perea-Moreno, M.-A., Samerón-Manzano, E. and Perea-Moreno, A.-J. (2019). Biomass as renewable energy: worldwide research trends. *Sustainability* 11: 863.

Petrescu, F. I., Apicella, A., Petrescu, R. V., Kozaitis, S., Bucinell, R., Aversa, R. and Abu-Lebdeh, T. (2016). Environmental protection through nuclear energy. *American Journal of Applied Sciences* 13: 941–946.

Priya, A. D. and Setty, Y. P. (2019). Cashew apple juice as substrate for microbial fuel cell. *Journal of Fuel Processing Technology* 246: 75–78.

Rachinski, S., Carubelli, A., Mangoni, A. P. and Mangrich, A. S. (2010). Microbial fuel cells used in the production of electricity from organic waste: a perspective of future. *Química Nova* 33: 1773–1778.

Ramseur, J. L. (2017). *US Carbon Dioxide Emissions Trends and Projections: Role of the Clean Power Plan and Other Factors*. Congressional Research Service, Washington, DC.

Rodionova, M. V., Poudyal, R. S., Tiwari, I., Voloshin, R. A., Zharmukhamedov, S. K., Nam, H. G., Zayadan, B. K., Bruce, B. D., Hou, H. and Allakhverdiev, S. I. (2017). Biofuel production: challenges and opportunities. *International Journal of Hydrogen Energy* 42: 8450–8461.

Romano, S. D. and Sorichetti, P. A. (2010). Introduction to biodiesel production. *Dielectric Spectroscopy in Biodiesel Production and Characterization*. Springer.

Sahu, B. K. (2018). Wind energy developments and policies in China: a short review. *Renewable Sustainable Energy Reviews* 81: 1393–1405.

Santoro, C., Arbizzani, C., Erable, B. and Ieropoulos, I. (2017). Microbial fuel cells: from fundamentals to applications. A review. *Journal of Power Sources* 356: 225–244.

Scarlat, N., Dallemand, J.-F. and Fahl, F. (2018). Biogas: developments and perspectives in Europe. *Renewable Energy* 129: 457–472.

Seddegh, S., Wang, X., Henderson, A. D. and Xing, Z. (2015). Solar domestic hot water systems using latent heat energy storage medium: a review. *Renewable Sustainable Energy Reviews* 49: 517–533.

Singh, V. K. and Singal, S. (2017). Operation of hydro power plants-a review. *Journal of Renewable Sustainable Energy Reviews* 69: 610–619.

Sitharthan, R., Swaminathan, J. and Parthasarathy, T. (2018) Exploration of wind energy in India: a short review. *National Power Engineering Conference (NPEC)*, 2018. IEEE, Madurai, India, pp. 1–5.

Sun, D. and Wang, L. (2016). Research on heat transfer performance of passive solar collector-storage wall system with phase change materials. *Energy Buildings* 119: 183–188.

Taromi, A. A. and Kaliaguine, S. (2018). Green diesel production via continuous hydrotreatment of triglycerides over mesostructured γ-alumina supported NiMo/CoMo catalysts. *Fuel Processing Technology* 171: 20–30.

Toklu, E. J. (2017). Biomass energy potential and utilization in Turkey. *Renewable Energy* 107: 235–244.

Ullah, K., Ahmad, M., Sharma, V. K., Lu, P., Harvey, A., Zafar, M., Sultana, S. and Anyanwu, C. (2014). Algal biomass as a global source of transport fuels: overview and development perspectives. *Progress in Natural Science: Materials International* 24: 329–339.

Von Hippel, F., Takubo, M. and Kang, J. (2019). The dream: a future powered by Plutonium. *Plutonium*. Springer, Singapore.

Weiland, P. (2010). Biogas production: current state and perspectives. *Applied Microbiology Biotechnology* 85: 849–860.

Zhang, B., Wang, B. and Wang, Z. (2017). Role of renewable energy and non-renewable energy consumption on EKC: evidence from Pakistan. *Journal of Cleaner Production* 156: 855–864.

Zou, C., Zhao, Q., Zhang, G. and Xiong, B. (2016). Energy revolution: from a fossil energy era to a new energy era. *Natural Gas Industry B* 3: 1–11.

2 Biomethanization in Anaerobic Systems
Feedstock and Process Enhancement

N. C. Ezebuiro
National Biotechnology Development Agency

C. O. Onyia
Godfrey Okoye University

I. Körner
Technische Universitaat Hamburg, Harburg

CONTENTS

DOI: 10.1201/9781003180289-2

2.1 INTRODUCTION

Anaerobic digestion (AD) has gained popularity in recent times as a technology for the addition of value to biomass and climate change mitigation due to its ability to recover methane from biomass. Methane recovery from biomass is essential because biogas, a product of AD, contains methane that could be used for electricity generation and as domestic cooking gas. With regard to climate change, methane is a potent greenhouse gas that must be captured and reused. Otherwise, when released into the atmosphere, methane causes global warming (and accelerates climate change) with its attendant health, environmental and socioeconomic effects. Hence, understanding and improving AD technology, especially methanization processes, at industrial scale, contributes in no small measure to capturing methane from biomass for reuse in electricity generation and heating, and mitigating climate change.

2.1.1 Fundamentals of Anaerobic Digestion

Municipalities, industries and agriculture are sources of biomass that could be used as feedstock in AD. These biomass feedstocks include food residues from restaurants; process residues from agro-industries; silages from grasses and agricultural residues; and wastewater. Other biomass feedstocks used in AD include energy crops such as maize silage and sunflower. Biomass feedstocks are composed of degradable and nondegradable components. The complex degradable organics are hydrolyzed to simpler compounds such as sugars, amino acids and fatty acid of various carbon lengths. The simpler compounds are further converted to energy-rich biogas through interdependent microbial processes (Schink and Stams, 2012). Biochemically nondegradable fractions remain as digester solids or digestate.

During stable biomass methanization, methane (CH_4) content of the biogas could range between 50% and 75% depending on whether carbohydrate- or lipid-rich substrates were used as feedstock. Other constituents of biogas could include CO_2 (25%–50%), O_2 (< 1%), and N_2 (<1%). Traces of hydrogen sulfide gas (H_2S), hydrogen gas (H_2) and chlorinated hydrocarbons could also be present, depending on the operating conditions (Appels et al., 2008).

2.1.2 Operating Conditions

Biomass could be fed to AD digester in batch, continuous or semicontinuous mode. AD temperature could be mesophilic or thermophilic, and a variety of digester designs are utilizable. For any biomass, AD temperature or digester design, optimum methanization is only possible with the right range of pH; C, N and S ratio; appropriate biomass sizes; and micronutrients (Gustavsson et al., 2011). Determining what ranges of the general factors that constitute optimum condition during methanization is a challenge. This is because biomass characteristics vary widely in composition, and different microorganisms are associated with different substrate types, digester design and temperature. Hence, methanization conditions are optimized for the microbes that produce CH_4 or for the microbes that degrade the fatty acids to CH_4-producing intermediates. Some of the conditions for optimum methanization are discussed next.

2.1.2.1 Temperature

Temperature stimulates changes in the composition, growth rate and metabolism of the microorganisms and also affects substrate degradation rates. Thermophilic temperature, T_t (45°C ≤ T_t ≤ 60°C), can speed up slow and energy-demanding reactions such as the degradation of volatile fatty acids (VFAs) to acetate, CO_2 and H_2. Conversely, mesophilic temperature, Tm (35°C ≤ Tm ≤ 45°C), is required for reactions which are exergonic (energy yielding), e.g., hydrogen-dependent methanogenesis (Kleerebezem and Stams, 2000). Temperature variation ≥ 1°C/day in both mesophilic and thermophilic AD is detrimental to optimum methanization, but variations ≤ 0.6°C/day are tolerable. Methanogens are the most sensitive to variations in temperature, and aceticlastic methanogens are more sensitive than hydrogenotrophs (Ahn and Forster, 2002).

2.1.2.2 Process Duration and Substrate Retention Time

Depending on substrate type, loading rate and other process conditions, substrate retention in the digester could range between 5 and 60 days. 5–10 days are recommended for microbial stabilization and substrate methanization in sewage sludge and biowaste. Conversely, longer retention time is required for lipid-like substrates such as grease trap residue, which could begin CH_4 formation not earlier than 10 days (Appels et al., 2008).

2.1.2.3 Volatile Fatty Acids/Alkalinity and pH

The ratio of VFAs to alkalinity during methanization is principally controlled by the concentrations of VFAs and hydrogen carbonate (HCO_3^-). Ammonia (NH_3) and ammonium (NH_4^+) are pH-dependent and interconvertible and could also contribute to alkalinity (Bardi and Aminirad, 2020). The ratio of VFAs to alkalinity is related to C:N since N content is related to NH_3 formation and VFAs derive from C content. Any ratio between 30 and 35 is optimum for maintaining optimum VFAs/alkalinity. For stable pH during AD, a steady HCO_3/VFAs molar ratio of 1.4 should be maintained. pH influences microbial rate of growth, metabolism and dominance, as well as composition of VFAs. For example, acetate-propionate dominance is established at pH 8.0 and acetate-butyrate dominance prevails at pH < 8. The phases of methanization also vary in optimum pH: a pH range of 4.0–8.5 is suggested for acidogenesis; and a narrower pH range of 6.5–7.4 is recommended for methanogenesis (Wainaina et al., 2019).

2.1.3 Phases of Anaerobic Digestion

AD proceeds in four biochemically interdependent phases. Each phase generates intermediate products which become substrates for succeeding stages until the biomass is completely mineralized. These phases include hydrolysis, acidogenesis, acetogenesis and methanogenesis.

2.1.3.1 Hydrolysis

Substrate hydrolysis is essentially a solubilization of macromolecules, and it is the slowest phase in methanization. High-molecular-weight compounds such as lipids, polysaccharides, proteins and nucleic acids are converted into soluble organics such as sugars, amino acids and fatty acids. Hydrolysis could be solely by physicochemical means or aided by enzymes of hydrolytic bacteria. The physicochemical factors include substrate composition and pH, and biochemical factors include the concentrations of the hydrolytic enzymes and enzyme-substrate interactions (Rajagopal and Béline, 2011).

2.1.3.2 Acidogenesis (Acidification)

Soluble organic intermediates derived from the biomass hydrolysis are split during acidogenesis to VFA molecules with carbon length between 1 and 5 (C1–5) by heterogeneous fermentative bacteria (acidogenic bacteria). Other products of this phase include alcohols (KOH), H_2S, NH_3 and CO_2. Acidogenic bacteria are both obligate and facultative anaerobes. During acidogenesis, factors such as pH, interspecies H_2 transfer and inoculum age play significant roles in converting soluble organics into VFAs (Mata-Alvarez, 2003).

2.1.3.3 Acetogenesis

In this phase, acetate-producing bacteria (acetogens or acetogenic bacteria) degrade the VFAs produced in acidogenesis to acetate, CO2 and H2 (Equations 2.1 and 2.2). The H2 concentration must be kept low to avoid inhibition. Therefore, most of the acetate is produced by syntrophic feeding relationship between VFAs-oxidizing bacteria and H_2-utilizing microorganisms (Scholten and Conrad, 2000). Certain species of the *Clostridium* and *Pseudomonas* genera are capable of producing acetate as sole product from CO_2 and H_2 using the reductive acetyl-CoA pathway as shown in Equation 2.3. These species are called homoacetogens, and the process is referred to as homoacetogenesis (Ragsdale and Pierce, 2008)

$$\text{Propionate} + 3H_2O \rightarrow \text{Acetate} + HCO_3^- + H^+ + 3H_2 \qquad \Delta G_0 = +76.1 \text{ kJ/mol} \quad (2.1)$$

$$\text{Butyrate} + 2H_2O \rightarrow 2\text{Acetate} + H^+ + 2H_2 \qquad \Delta G_0 = +48.3 \text{ kJ/mol} \quad (2.2)$$

$$4H_2 + 2HCO_3^- + H^+ \rightarrow \text{Acetate} + 4H_2O \qquad \Delta G_0 = -104.6 \text{ kJ/mol} \quad (2.3)$$

2.1.3.4 Methanogenesis

Methanogenesis completes the mineralization of biomass to CO_2 and CH_4. Important biomass intermediates for methanogenesis include acetate, CO_2, H_2, and methylated compounds such as methanol, methylamine and dimethylsulfide. Other biomass intermediates include carbon monoxide (CO) and formate ($HCOO^-$). Methanogenesis complements acetogenesis by the utilization of acetogenic substrates such as formate and H_2, thereby making VFAs degradation energetically favorable and thermodynamically feasible (Schink and Stams, 2012). Figure 2.1 summarizes the flow of carbon intermediates between the different phases of methanization. Part of the energy generated from degrading the substrates is retained in the C-H bond in CH_4 and is not available to the microorganisms involved in AD. The retained energy is the important resource in CH_4. Retaining the energy of degradation of VFAs in CH_4 is also responsible for lower energetics and slower degradation of substrates during AD when compared with aerobic digestion (Kleerebezem and Stams, 2000).

2.1.4 Biomass Degradation to Volatile Fatty Acids

VFAs are produced during acidogenesis as degradation products of long-chain fatty acids (LCFA). The carbon length of short chain VFAs could range between 1 and 5, and these include formate, acetate, propionate and butyrate. Others VFAs encountered in AD include valerate and the isoforms of butyrate and valerate. The most

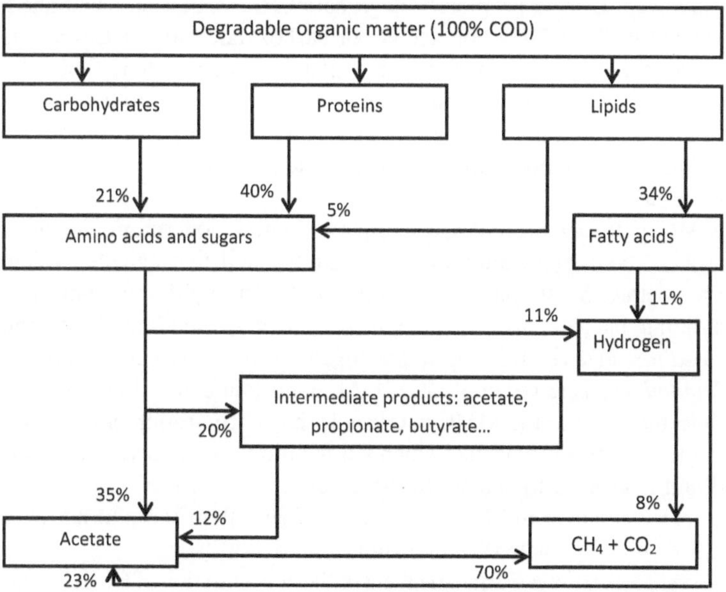

FIGURE 2.1 Bioconversion of substrates in anaerobic digestion with flow of substrate intermediates expressed as percent of chemical oxygen demand (COD). (Modified by Ezebuiro, (2014) from Siegrist et al., 1993.)

important VFAs are acetate, propionate and butyrate because they are the regular intermediate products from the beta-oxidation of LCFA. Small amount of propionate is produced by beta-oxidation of LCFA with odd number of carbon atoms. Oxidative degradation of branched-chain amino acids (valine, isoleucine, threonine and methionine) also produces propionate. Butyrate is produced from degradation of LCFA with even number of carbon and from carboxylation of propionate. Acetate is a common intermediate in the degradation of VFAs (Meegoda et al., 2018). VFAs are activated and degraded to CH_4 and CO_2 by different processes and pathways that form the focus of the next subsection.

2.1.4.1 VFAs Activation

VFAs metabolism to intermediates that are utilizable by CH_4-forming microorganisms begins with activation. Activation requires biochemical energy and involves transfer of phosphate group (PO_3^{4-}) from adenosyl triphosphate (ATP) to the VFA to form phosphorylated VFA (acyl-phosphate) as shown in Equations 2.4 and 2.5:

$$\text{VFA} + \text{ATP} \rightarrow \text{Acyl} - \text{Pi} \tag{2.4}$$

$$\text{Acyl} - \text{Pi} + \text{SH} - \text{CoA} \rightarrow \text{Acyl} - \text{CoA} + \text{ATP} \tag{2.5}$$

The phosphorylated VFA further reacts with coenzyme A (SH-CoA) to form an acyl-CoA. The activated VFA enters the appropriate degradation pathway depending on the VFA involved. During propionate degradation, the 3-Carbon backbone is either decarboxylated to form 2-Carbon acetate or carboxylated to a 4-Carbon molecule that can undergo beta-oxidation. Nutrient availability, process conditions and microbial composition are the major determinants of the prevalent pathway for propionate degradation during methanization (Ziemiński and Frac, 2012). Two important pathways that are commonly reported include the methylmalonyl-CoA (MMC) and butyryl-CoA pathways.

The MMC pathway produces a complete degradation of propionate to CH_4. MMC involves the carboxylation of propionate to the 4-C succinate, which is beta oxidized to acetate. MMC pathway is common to syntrophic relationship between VFAs-oxidizing bacteria species of *Syntrophobacter*, including *S. pfennigii* and *S. fumaroxidans*, and H_2-utilizing methanogens of the orders *Methanobacteriales* and *Methanomicrobiales* (Li et al., 2012). An important determinant of the acetate-yielding intermediates in the MMC pathway is the concentration of Co. Succinate is produced and converted to acetate when Co is optimum, and hydroxypropionate is produced and converted to acetate in the absence of Co. MMC pathway involving succinate intermediate could be more energy saving than the alternative pathway involving hydroxypropionate (Scholten and Conrad, 2000).

The butyryl-CoA pathway is independent of the concentration of Co. It is found in the syntrophic relationships between members of the genera *Smithella* and *Syntrophomonas*, and aceticlastic methanogens of the genera *Methanosaeta* and *Methanosarcina*. It involves conversion of propionate to hydroxyl-butyryl-CoA and β-oxidation of hydroxyl-butyryl-CoA to acetate (Kleerebezem and Stams, 2000).

2.1.4.2 VFAs Accumulation and Anaerobic Digester Failure

Accumulation of VFAs inhibits further biomass degradation and induces digester failure. H_2, formate and acetate are important intermediates generated from long-chain VFAs oxidation that must be removed or kept at very low concentrations for VFAs degradation to be thermodynamically feasible. The removal of VFAs degradation intermediates controls VFAs degradation rate and bioenergetics. The accumulation of H_2 or formate in AD reactors inhibits acetate, propionate and butyrate oxidation, whereas the accumulation of H_2, formate, acetate or butyrate inhibits propionate oxidation (Li et al., 2012). Furthermore, reactor acidification, especially due to propionic acid accumulation during methanization, induces loss of HCO_3^- buffer as gaseous CO_2, and this partly accounts for high CO_2 content in biogas during conditions of failed methanization (Wainaina et al., 2019).

2.1.4.2.1 Suppression of Methanogens as Cause of Digester Failure

Suppression of methanogens that utilize the intermediate products of VFAs degradation results in VFAs accumulation and digester failure. Suppression of methanogens could be due to presence of inhibitory substances or as a result of competition for H_2 by sulfate-reducing bacteria (SRB). It could also be as a result of process modification to generate and recover VFAs as industrial raw materials. Methanogenic suppression is generally accompanied by low pH and decline in CH_4 content in biogas. Accumulation of propionic acid and, to a lesser extent, the iso-forms of valeric and butyric acids generally precedes methanogenic suppression (Magdalena et al., 2019).

2.1.4.2.2 Dominance of Unionized VFAs Species as Cause of Digester Failure

Digester failure is more often associated with conditions that induce accumulation of unionized forms of VFAs, and the most important condition is pH. pH ranges that keep the VFAs in ionized state reduce the risk of digester failure in spite of relatively high VFAs concentrations. A pH range of 7.8–8.2 for AD is recommended to maintain VFAs in ionized forms and eliminate SRB competition. Methanization pH is particularly important for propionic acid because it is usually in the highest concentration when VFAs accumulate, and 7.0–8.5 is an appropriate methanization pH range for keeping propionic acid concentration low and degradable (Li et al., 2012). It has been established that accumulation of propionic acid is related to the following biochemical conditions:

- Similarity in the enzymes (Kinases) of propionate and acetate degradation, with acetate-Kinase having better kinetics (Ingram-Smith et al., 2005); and
- Joint inhibition by butyric- and acetic acids, with butyric acid being about 2.4 times more inhibitory to propionic acid degradation than acetic acid (Amani et al., 2010).

2.1.4.3 Monitoring VFAs Accumulation and Digester Failure

Deviation from regularly observed VFAs levels or established VFAs degradation rates and CH_4 yields is the most reliable indicator that the microorganisms are responding to inhibitory influences. However, there is no consensus on VFAs concentrations that constitute the "critical levels." It is suggested that VFAs concentration ≥ 50 mmol/L is ideal for stable methanization, and ≥ 200 mmol/L could induce instability in AD (Batstone et al., 2000; Voss et al., 2009; Zhang et al., 2010). Fatty acid/alkalinity (FOS/TAC) is also an important parameter for monitoring process stability during methanization. An FOS/TAC value of 0.15–0.45 indicates stability, whereas values above 0.6 are an indication of the onset of instability (Voss et al., 2009).

2.1.4.4 Other Biochemical Factors That Affect VFAs Degradation

Factors that affect the growth of the desired microbial population are capable of impeding VFAs utilization. These are discussed.

2.1.4.4.1 Methanogen and Acetogen Ratio (M/A)

The methanogen and acetogen ratio (M/A) of the microbial population participating in syntrophic feeding relationship is an important parameter for VFAs degradation. Considering that methanogens are slow growing anaerobes, relatively high proportion of methanogens is required at the start-up of methanization compared to fast growing acetogens. In a study, Amani et al. (2011) indicated that utilization efficiency for propionic, butyric and acetic acids increased from 10% to almost 60% by increasing M/A from 1:1 to 2:1. Conversely, at M/A of 3:1, the advantage for VFAs degradation was lost, acidification ensued and the accompanying low pH resulted in acetate accumulation. This was because there were less propionic- and butyric acid oxidizing bacteria than were required to carry out degradation of the accumulating VFAs to acetate.

2.1.4.4.2 Inoculum Structure

VFAs must be in contact with microorganism or with microbial enzymes in the reactor in order to be degraded. This requires that the diffusion distance between the microorganisms and the substrates must be short. Short diffusion distance allows for rapid transfer of metabolites and increases degradation energetics and kinetics. Short diffusion distance exists between VFAs and microorganism when the microorganisms are in clusters (e.g. biofilms). This is particularly important for H_2 and formate metabolism (Kleerebezem and Stams, 2000).

2.1.4.4.3 Microbial Competition for Carbon Sources

Most competition during methanization is between microorganisms that use H_2 generated during VFAs degradation to reduce CO_2 to CH_4 (MPB) and microbes that reduce SO_4^{2-} to H_2S (SRB). SRB possess stronger kinetic affinity for H_2 than MPB. MPB produce CH_4 (Equation 2.6), and SRB produce H_2S (Equation 2.7) with H_2 generated during VFAs degradation. The use of H_2 by SRB to produce H_2S is at the expense of methanogenesis and results in increased concentration of H_2S in biogas and toxicity of sulfide ion (HS^-) to MPB (Chen et al., 2008).

$$4H_2 + HCO_3^- + H^+ \rightarrow CH_4 + 3H_2O \qquad \Delta G_0 = -135.5 \text{ kJ/mol} \qquad (2.6)$$

$$4H_2 + SO_4^{2-} + H^+ \rightarrow HS^- + 4H_2O \qquad \Delta G_0 = -151.9 \text{ kJ/mol} \qquad (2.7)$$

Just like methanogens and acetogens, SRB grow on propionate, butyrate and acetate. SRB have higher affinity for and derive more energy from H_2 and propionate compared to acetate and butyrate. Equations 2.8 and 2.9 show the energetics advantage of SRB over acetogens respectively in propionate conversion to acetate.

$$\text{Propionate} + 0.75SO_4^{2-} \rightarrow \text{acetate}$$

$$+ HCO_3^- + 0.75HS^- + 0.25H^+ \qquad \Delta G_0 = -37.7 \text{ kJ/mol} \tag{2.8}$$

$$\text{Propionate} + 2H_2O \rightarrow \text{acetate} + CO_2 + 3H_2 \qquad \Delta G_0 = +76.1 \text{ kJ/mol} \tag{2.9}$$

Identifying factors that enable SRB to outcompete other microbes of methanogenic significance is important for methanogenesis. Two factors have been widely reported and these include Chemical Oxygen Demand/SO_4^{2-} and SRB/MPB; and temperature. Chemical Oxygen Demand (COD) and SO_4^{2-} ratio and the start-up ratio of SRB and MPB influence aceticlastic methanogenesis. Operational AD temperature also plays a significant role during H_2 utilization by SRB and MPB. Colleran and Pender (2002) reported that H_2-dependent reduction of SO_4^{2-} prevails over H_2-dependent CH_4 formation at 35°C–37°C; conversely, H_2-dependent CH_4 formation prevails over H_2-dependent SO_4^{2-} reduction at 55°C. This might suggest that processes involving MBP are reasonably optimized by temperature in thermophilic methanization, while mesophilic methanization requires optimization measures for MPB for enhanced H_2 utilization.

2.1.4.4.4 Inhibitory Substances

A substrate or its intermediate is inhibitory when it affects growth of microbes, adversely changes the microbial population, kinetics or energetics of a particular phase with reference to the generation of a choice product. NH_3, SO_4^{2-} and HS are dominant inhibitory intermediate products of substrate degradation that lead to reduction in the rate of CH_4 production and cause accumulation of VFAs (Chen et al., 2008). NH_3 is produced from degradation of urea and protein-rich compounds. NH_3 is protonated to ammonium ion (NH_4^+) in low pH solution, and both species form the total ammonia nitrogen (TAN) (Meegoda et al., 2018). NH_3 is capable of diffusing across bacterial and archaeal membranes and causing adjustment in intracellular proton balance. Other toxic intracellular effects of NH_3 include interferences with process bioenergetics and enzyme inhibition (Gallert and Winter, 2008).

NH_3 toxicity causes digester instability due to the accompanying fluctuations in process pH. The ratio of NH_3/NH_4^+ is dependent on process temperature and pH, and total ammonia nitrogen $\leq$ 5.0 g/L with an NH_3 concentration of about 1.0 g/L (pH 7.9; 55°C) is toxic to both acetate-producing and acetate-utilizing microorganisms. Higher temperature generally increases toxicity of NH_3, but this can be attenuated by adding Cu and Co to anaerobic digesters (Bardi and Aminirad, 2020).

2.1.5 Enhancing Methanization through Metal-Based Biocatalysis

Generally, substantial amount of the energy generated during methanization is trapped in CH_4 and is unavailable for biochemical work. Biocatalysis has evolved in AD to counteract the low energy exchange associated with methanization in general and

VFAs degradation in particular. Low AD energetics implies that some of the AD reactions cannot go on spontaneously and must be catalyzed. The incorporation of trace elements (TEs) into the enzymes of microorganisms that are involved in AD reactions with low energy exchange ensures that the rate of these reactions meet thermodynamic requirements. Enhancing the activities of these metal-containing enzymes or metalloenzymes (MEs) that are associated with AD is an important aspect in methanization process enhancement and is referred to as biocatalysis. Different phases of AD have different MEs that perform varying functions. These are discussed.

2.1.5.1 TEs and MEs of Hydrolysis

Hydrolysis is a solubilization reaction of the type shown in Equation 10 in which R is an organic compound, and the transfer of the OH^- from H_2O in an aqueous solution could be influenced by pH, a catalyst, an enzyme or all three.

$$RX + H_2O \rightarrow ROH + HX \tag{2.10}$$

Hydrolytic enzymes are called hydrolases and could hydrolyze AD substrates inside (intracellular) or outside (extracellular) the cell of the microorganism. Hydrolases could also be both intra- and extracellular depending on the type of reaction that is catalyzed. However, the important hydrolases of AD are extracellular enzymes and include lipases, glucosidases and proteases (Burgess and Pletschke, 2008).

The active sites of hydrolases generally consist of charged amino acid residues but might also contain one or two divalent metal ions such as Ca^{2+}, Mg^{2+}, Zn^{2+} and Co^{2+} (Heikinheimo et al., 2001). The hydrolytic potential of the MEs varies with the number of coordinating divalent metal ions, and the hydrolytic effect of two divalent metal ions is twice as high as that of a single divalent metal ion. This explains the presence of two divalent metals ions participating in the hydrolytic mechanisms of MEs such as ureases, alkaline phosphatase and inorganic pyrophosphates among others.

2.1.5.2 TEs and MEs of Acetogenesis

The predominant pathway for growth and metabolism in acetogens is the acetyl-CoA pathway. Acetyl-CoA pathway involves metabolic processes that reduce CO_2 or CO to form acetate by using H_2 as electron donor. In the reverse direction, acetate is oxidized to its precursors. Acetyl-CoA pathway is present in both acetogens and methanogens involved in the conversion of acetate to CH_4 and reduction of CO_2 to acetate using H_2 (Drake et al., 2008). The principal enzymes of the acetogenic and methanogenic acetyl-CoA pathway are MEs and include hydrogenases, FDH, MeTr and CODH/ACS.

2.1.5.2.1 Hydrogenases

During CO_2 reduction to methane, hydrogenases provide redox electrons by oxidation of H_2. Some hydrogenases contain metals, while others do not. The metals of importance in hydrogenases include Ni and Fe (Figure 2.2a); other components include amino acid residues and sulfur (S). Ni-Fe complex is necessary for substrate binding, and Fe-S complex (Figure 2.2b) is required as conduit for electron transfer (Shima et al., 2002).

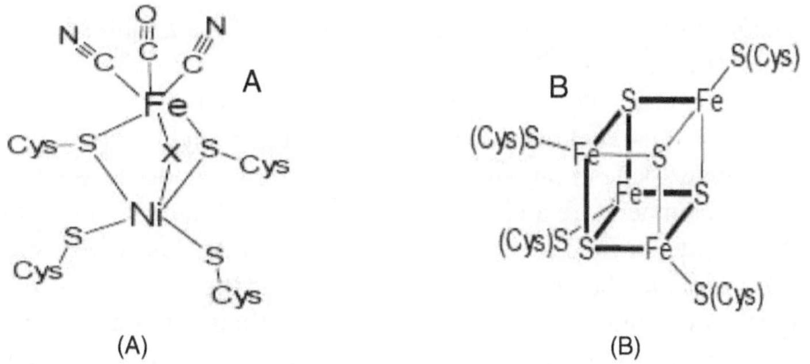

FIGURE 2.2 Ni-Fe catalytic center in hydrogenases (a); Fe-S cluster in hydrogenases (b) (Shima et al., 2002).

2.1.5.2.2 Formate Dehydrogenases (FDH)

These MEs contains Se, W or Mo at the active site, and the TEs found at the active site depend on the microbial species, the growth substrates or bioavailability of the TEs in the growth medium. FDH catalyzes thermodynamically unfavorable CO_2 reduction to formic acid (HCOOH). Redox electrons for the biocatalysis are obtained from reduced nicotinamide adenine dinucleotide phosphate (NADPH), H_2 or pyruvate. When CO is the substrate instead of CO_2, FDH is associated with oxidation of CO to CO_2 and formic acid (Reda et al., 2008; Seravalli and Ragsdale, 2000).

2.1.5.2.3 Methyltransferase (MeTr)

MeTr is associated with transfer of $-CH_3$ and is unique to microorganisms that employ the acetyl-CoA pathway for acetate metabolism. It has either of two types of active-site complexes: cobalamin or cobamide, and both are activated by Co. The active site of MeTr is capable of accepting $-CH_3$ groups when the Co is in Co^+ state to form an enzyme-bound CH_3-Co^{3+} complex. The $-CH_3$ group is transferred to coenzyme M (SH-CoM) to form CH_3-S-CoM, and the active Co^+ state is regenerated. Temperature range between 35°C and 40°C and pH range between 7.5 and 7.7 are optimum for MeTr-dependent biocatalysis. The enzyme could be stable at temperatures up to 45°C, and activity-decline sets in around 70°C (Caspi et al., 2008).

2.1.5.2.4 CODH/ACS Complex

CODH/ACS complex is the main enzyme for the metabolism of acetate in both acetogens and methanogens. It comprises two sub-units: CODH and ACS. CODH/ACS complex is involved in the metabolism of H_2, CO_2, CO and VFAs. During CO_2 metabolism, one molecule of CO_2 is reduced to CO by CODH, and this becomes the carbonyl group of acetate. Another molecule of CO_2 is reduced to formate ($HCOO^-$) and further to $-CH_3$ group, and this forms the $-CH_3$ group of acetate. The ACS unit condenses the CO and $-CH_3$ with CoA to form acetyl-CoA. Structurally, the active sites of the CODH/ACS complex in anaerobes contain Ni ions linked to the Fe-S electron conduit (Lindahl, 2004).

CODH/ACS complex may additionally contain Mo and Cu to detoxify reactive oxygen species (Gnida et al., 2003). The biochemical and kinetic properties of CODH/ACS complex depend on whether it is involved in acetogenesis or methanogenesis. Optimum temperature and pH for CODH are 60°C and 6.3, respectively, for acetogenic CO_2 reduction, and about 70°C is optimum for ACS in aceticlastic methanogenesis. This explains the observation that in extreme thermophilic AD (≥60°C), CO_2 and H_2 are converted to acetate instead of CH_4 (Iranpour et al., 2005). It also explains reports that CH_4 production in thermophilic AD is optimum compared to mesophilic AD that require TEs supplementation for optimum CH_4 production (Uemura, 2010).

2.1.5.3 TEs and MEs of Methanogenesis

CH_4 production is possible through three different routes by different methanogens, and these include hydrogenotrophic methanogenesis where CH_4 is formed from H_2 reduction of CO_2. Others include aceticlastic methanogenesis involving the cleavage of acetate into CO_2 and CH_4 and methylotrophic methanogenesis comprising redox reactions using methylated substrates such as methanol, methylamines and dimethylsulfide to form CH_4. Some MEs that are common to both acetogenesis and methanogenesis such as MeTr and CODH/ACS have been discussed. However, MEs associated with CH_4 formation from acetate or H_2 reduction of CO_2 are unique to methanogens (and to some extent, are also involved in sulfate reduction). These unique MEs of methanogenesis and their TEs relations are discussed next.

2.1.5.3.1 Formyl-Methanofuran Dehydrogenase (FMD)

FMD is the methanogenic analog of FDH. It is also called methanofuran reductase (MR) (Ragsdale and Pierce, 2008). Some important features of FMD include: the presence of Mo or W at the active site (depending on which of the TEs is bio-available in the substrates), but it is more active with Mo than with W.

2.1.5.3.2 Methyl-CoM Reductase (MCR)

MCR is only found in methanogens and catalyzes the final stage of CH_4 production from the reduction of $-CH_3$. This involves the reaction between methyl-coenzyme-M (CH_3-S-CoM) and coenzyme-B (SH-CoB) in the presence of MCR to form CH_4, SH-CoM and SH-CoB. Two forms of MCR known as MCR-I and MCR-II exist: MCR-I functions optimally in the pH range between 7.0 and 7.5 and has a maximum reaction rate of approximately 6 μmol/min/mg. The maximum reaction rate for MCR-II is approximately 21 μmol/min/mg and the optimum pH range is 7.5–8.0. Notwithstanding the form of existence, the general feature of MCR includes a reactive co-factor known as F_{430}, which is dependent on its active Ni-complex for catalysis. Ni-complex in MCR is similar to the Co-complex in MeTr, and the active form of Ni in MCR is the Ni^+ state, which forms CH_3-Ni^{3+} when it receives CH_3 group from MeTr (Caspi et al., 2008).

2.2 CONSIDERATIONS FOR THE USE AND APPLICATIONS OF METAL-BASED BIOCATALYSIS IN AD PROCESSES

2.2.1 Considerations for Use of Metal-Based Biocatalysis in AD Processes

The biocatalytic potentials of TEs have resulted in the deployment of the technology in many aspects of AD. The success of such deployment depends on certain considerations. The three most important considerations include inoculum composition and feeding mode, composition of the biomass feedstock and interaction effects of the TEs with biomass VFAs.

2.2.1.1 Inoculum Composition and Feeding Mode

The efficiency or otherwise of TEs-based biocatalysis is influenced by the concentration of TEs in the biomass and in the inoculum (seeding sludge) used. The inoculum is a significant reservoir of TEs, and the composition of the inoculum is strongly influenced by the digested biomass and its source. Fe-rich sludge used as inoculum may result in limited AD improvement or even a decline in process stability and methane production relative to a control system compared to Fe-limited inoculum (Yazdanpanah et al. 2018). Figure 2.3 shows the TEs composition of inoculums from digested sewage sludge obtained from two sources and digested maize silage-based feedstock.

The TEs contents of the DS-1 and DS-2 varied widely, and this could be attributed to the different sources of the sludge and the digestion process. The MSF-D in Figure 2.3 has significantly elevated TEs contents because TEs were supplemented to the digester during the AD process. Continuous feeding operation in AD, which is usually accompanied by loss in seeding sludge, may result in reduced CH_4 yield due to loss of TEs during digestate removal. Conversely, batch feeding operations have the advantage that the starting TEs concentration and configuration remain in the

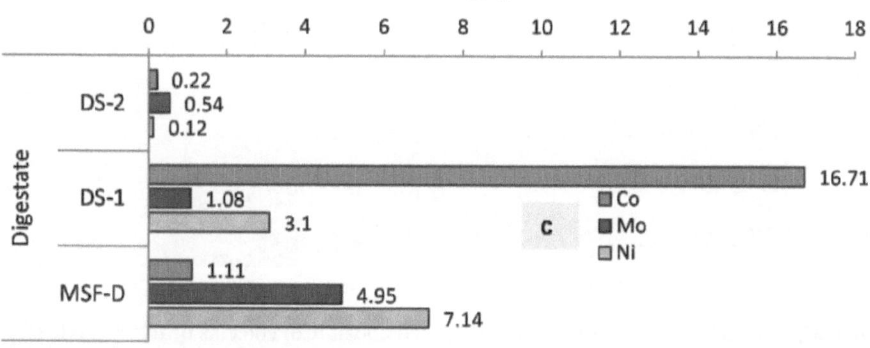

FIGURE 2.3 Nickel, molybdenum and cobalt contents of digested sewage sludge (DS-1 and DS-2: digested sludge 1 and 2; MSF-D, digestate from maize silage-based feedstock). (Adapted from Ezebuiro and Körner, 2017.)

digester all through the methanization period. Hence, even if the TEs are inadequate, this is not compounded by removal of biomass and inoculum during feeding.

2.2.1.2 Composition of the Biomass Feedstock

Stimulatory influences of TEs have been reported for AD from both mono-substrates and complex substrates. However, the degree of stimulation is directly related to the inherent concentration of TEs in the biomass. Figure 2.4 shows the Ni, Co and Mo composition of AD substrates based on biomass complexity. In many instances, Se concentrations in biomass are so low that they occur below the detection limit of the measuring equipment (Ariunbaatar et al., 2016; Ezebuiro and Körner, 2017).

Simple substrates predominantly comprise a biomass type such as grass specie or leaf type or a digested form of it as in the dung. In comparison to simple substrates such as LG and SG or ML (Figure 2.4a) that may constitute the original feedstock, dungs are generally known to have lower TEs contents. This could be due to assimilation of the nutrients by the animals for growth and metabolism. Hence, TEs supplementation is often beneficial for digesters using simple substrate during AD. Complex substrates may combine two or more simple substrates and, as a result, may have better TEs composition and configuration.

In Figure 2.4b, the complex substrates contain more TEs and reflect more variation in the TEs. For example, Ni content varies widely between 0.12 and 3.62 mg/kg DM compared to Mo and Co, and the MSF has the highest content of Ni, Co and Mo among the complex substrates shown. Such wide range of TEs in substrates are common place in reports relating to AD substrate characteristics (Hullebusch et al., 2005; Zandvoort et al., 2003). The implication of such variation in TEs is that specific TEs may need to be supplemented to produce a more balanced TEs mixture because they are insufficient in the biomass.

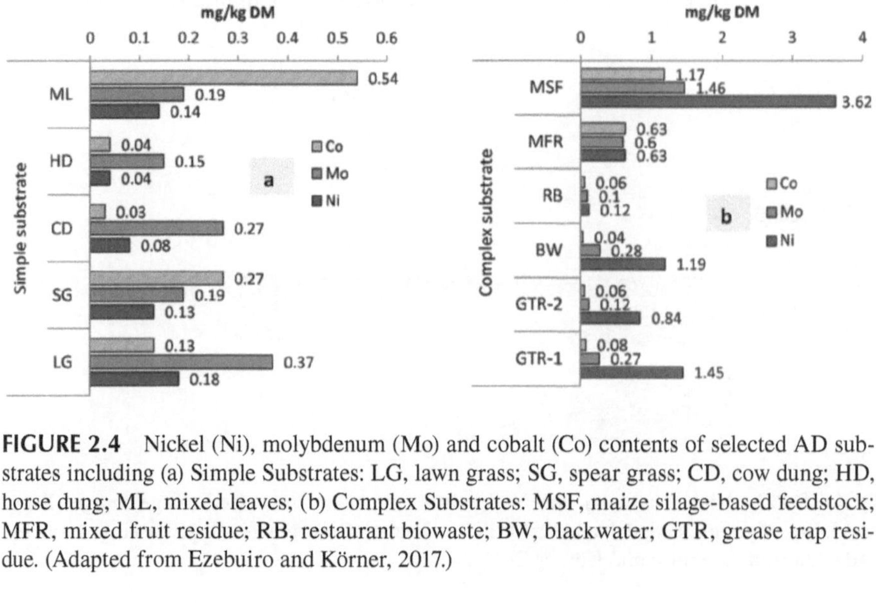

FIGURE 2.4 Nickel (Ni), molybdenum (Mo) and cobalt (Co) contents of selected AD substrates including (a) Simple Substrates: LG, lawn grass; SG, spear grass; CD, cow dung; HD, horse dung; ML, mixed leaves; (b) Complex Substrates: MSF, maize silage-based feedstock; MFR, mixed fruit residue; RB, restaurant biowaste; BW, blackwater; GTR, grease trap residue. (Adapted from Ezebuiro and Körner, 2017.)

2.2.1.3 Interaction Effects of TEs with VFAs from Biomass Degradation

The biochemical efficiency of TEs is enhanced when an individual TE is in interaction with another TE (TE*TE) or biomass degradation intermediate such as VFAs (VFAs*TE) (Ezebuiro, 2018; Higuchi et al., 2005). Generally, when two or more TE ions interact in a mixture, the individual influences of one or all the interacting ions could weaken (antagonism) or strengthen (synergism) the AD process under consideration. TEs mixtures of Zn, Cu and Ni; Ni and Cu; and Ni, Mo and Co have been established as synergistic in AD (Bardi and Aminirad, 2020; Ezebuiro and Körner, 2017). Regarding VFAs and TEs interactions and their effects, digester VFAs level is a significant consideration for the effects of TEs on methane production processes and VFAs metabolism. Figure 2.5a shows the significant TE*TE and VFAs*TE for

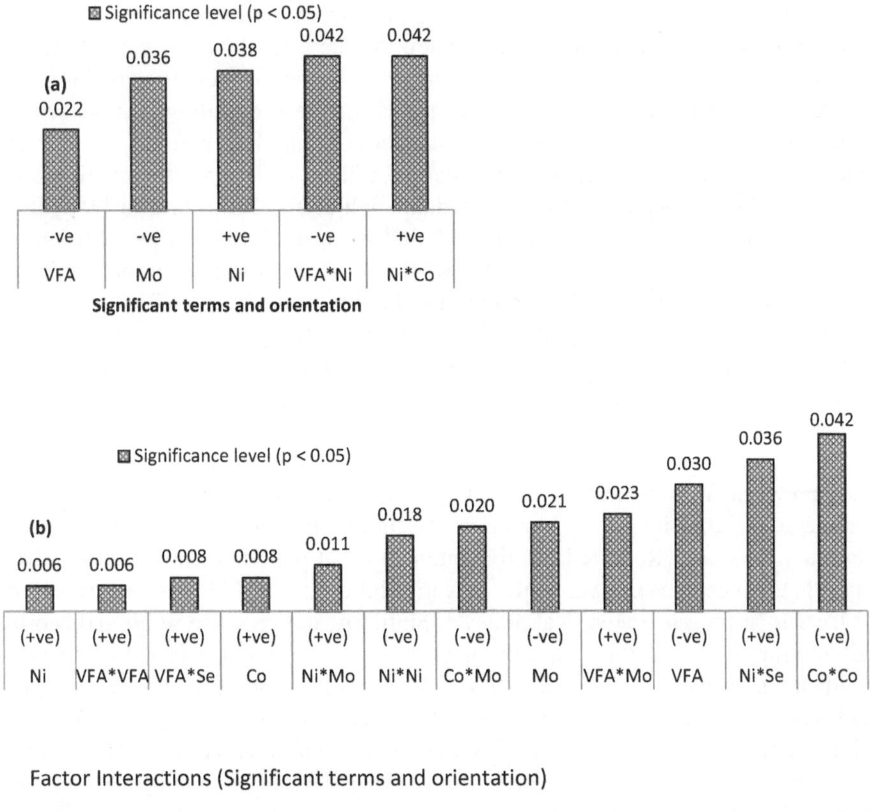

FIGURE 2.5 Significant TEs and VFAs interactions that influence (a) methane production. (b) VFAs degradation rate due to Ni, Co, Se and Mo supplementation to VFAs levels between 28 and 213 mmol/L at 37°C (Adapted from Ezebuiro et al., 2018). NOTE for Figure 2.5a and b: The factor is either an individual factor concentration (Ni, Co, Se, Mo or VFAs); or a combination of any of the individual factors (e.g. Ni*Mo, VFA*TE). The effect of a model term (also called factor effect) on a response may be positive (+ve) or negative (-ve) and the significance of the effect is given by the *p values*. The effect (+ve or –ve) of a model term on a response is significant when its *p value* is less than 0.05.

methane production, and Figure 2.5b shows for VFAs degradation rate due to Ni, Co, Se and Mo supplementation to VFA levels between 28 and 213 mmol/L at 37°C.

2.2.2 Applications of Metal-Based Biocatalysis in AD Processes

The most common applications of TEs in AD operations are in the improvement of hydrolysis and acidification to increase VFAs yield, increase in conversion of short chain fatty acids to acetate, and enhancement of process stability and production of methane from biomass.

2.2.2.1 Improvement of Hydrolysis and Acidification Rate

TEs supplementation has been shown to increase biomass hydrolysis and acidification. Hydrolysis is a limiting step during methanization in AD. The amount of VFAs available for metabolism is dependent on the rate of hydrolysis and acidification. Hence, increasing hydrolysis and acid formation potential of biomass ensures availability of intermediate substrates for the succeeding AD phases of acetogenesis and methanogenesis. Fe, Mo and Co supplementations have been specifically identified as having stimulatory effects on hydrolysis in digesters converting food waste to methane (Yazdanpanah et al., 2018). Similarly, hydrolysis and acidification of waste activated sludge can be improved by addition of potassium ferrate (He et al., 2018). HAR improvement is important not only for methanization but also for the targeted production of short chain VFAs such as acetate, propionate and butyrate. These intermediates form the precursors for many industrial products and have received increased attention lately with regard to AD process manipulations for enhanced production (Magdalena et al., 2019; Wainaina et al., 2019).

It is important to highlight that during hydrolysis, the accumulation of organic acids prevents further biomass hydrolysis. This challenge has been shown to be eliminated by exploiting the interaction effects of TEs and relationships between TE and VFA levels. Reported positive interactions at pre-acidification levels $\geq$200 mmol/L VFAs include Co*Se, while a negative influence on HAR has been reported for Co*Mo at the same acidification level. Similarly, the presence of Ni and Mo in TEs mixture and VFA*Mo improves hydrolysis of maize silage feedstock at pre-acidification levels of 10 and 120 mmol/L VFAs. It has also been shown that optimum concentrations and configurations of TEs are required for HAR at different acidification levels. Figure 2.6a–c shows optimum configuration of Ni, Co, Se and Mo supplementation for the enhancement of the HAR of maize silage biomass at different levels of pre-acidification during methanization. The process efficiencies at 10, 120 and 200 mmol/L VFAs pre-acidification levels as expressed by desirability function are 87%, 43% and 88%. Similarly, the relative HARs (compared to a baseline condition of no TEs supplementation) at the corresponding pre-acidifications levels are 82%, 13% and 85%. Such advanced analysis enables the computation of specific TEs requirements for the enhancement of biomass HAR at different levels of process acidification. This is particularly important considering that acid accumulation prevents biomass hydrolysis.

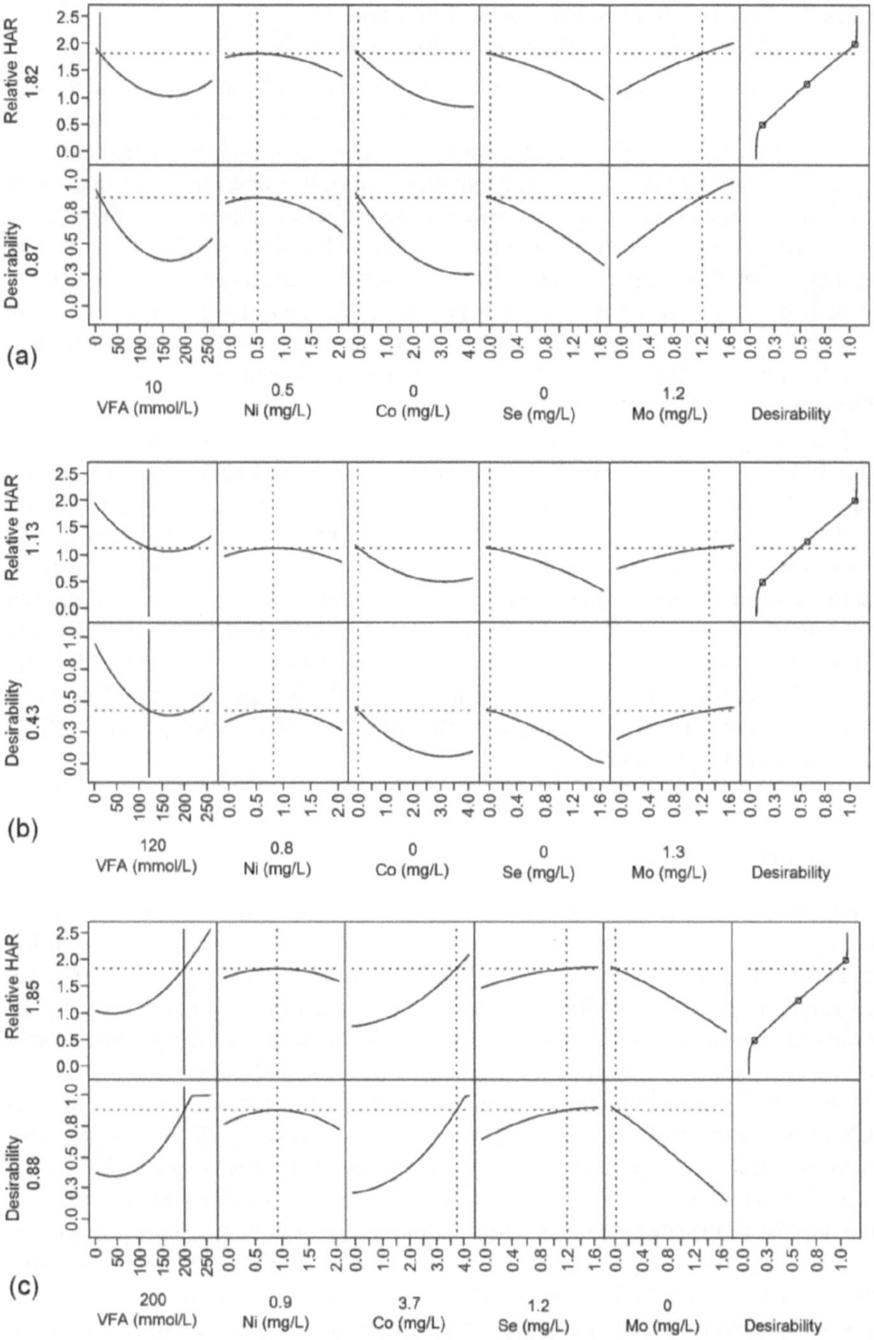

FIGURE 2.6 Optimum Ni, Co, Se and Mo configuration for the enhancement of hydrolysis and acidification rate in maize silage-based feedstock at a pre-acidification level of (a) 10 mmol/L; (b) 120 mmol/L; and (c) 200 mmol/L VFA operated at 37°C. (Adapted from Ezebuiro and Körner, 2017.)

2.2.2.2 Enhancement of Process Stability and Production of Methane from Biomass

Over the years, researchers have improved methane production by supplementing TEs. In various laboratory and pilot studies, methane production has been increased by up to 80% from baseline production figures as a result of TEs supplementation. In such systems, methane production enhancements are as a result of improved competition of methane producing archaea without affecting archaeal diversity (Chan et al., 2019). AD digesters using simple substrates such as silages from grasses benefit strongly from TEs supplementation because such biomass substrates are known to be deficient in TEs and have poor TEs balance in the digestion mixture (Pobeheim et al., 2010). Other methanization processes such as process stability and VFAs degradation rate have been reported to benefit from the biocatalytic potentials of TEs (Ortner et al., 2014).

Improving methane production based on TEs supplementation is sensitive to organic loading rate and VFAs concentration of the methanization digester. Figure 2.7 shows the optimum TEs configurations for methane production and the associated process efficiencies (as depicted by response surface modeling and desirability function) for different concentrations of digester VFAs. The percent increase in methane production (expressed as relative methane production) and corresponding process efficiency (expressed as desirability) at the shown TEs configuration considered to be optimum for a VFAs concentration include: 55% and 99% at 23 mmol/L VFAs (low organic loading rate); 32% and 89% at 125 mmol/L VFAs (medium organic loading rate); and 19% and 73% at 200 mmol/L VFAs (high organic loading rate).

2.3 CONCLUSION

Biomethanization from biomass is an anaerobic process that involves many stages and requires the understanding of many factors for optimum performance of the process. The need to understand the biochemical basis of the competition for intermediate products of biomass degradation such as VFAs has led to probes into biomass characteristics and biocatalysis. While biomass differs considerably in characteristics, biocatalysis for biomethanization enables the speeding of biochemical processes of biomass degradation to maximize methane yield. A special field of biocatalysis that has gained attention over the years is TEs-based enhancement of metalloenzymes-mediated AD processes. With this advanced response-specific enhancement technique, optimum TEs configurations could be developed and applied to targeted AD responses with increased efficiency. This precise science can be directly applied to the bioenergy industry to increase biomethane production from biomass and as well increase the share of electricity from biomass. It is anticipated that further research into the development of TEs formulation for specific bioenergy processes will further enhance the economics and spread of biomass methanization, especially in regions where conventional energy supply is inadequate.

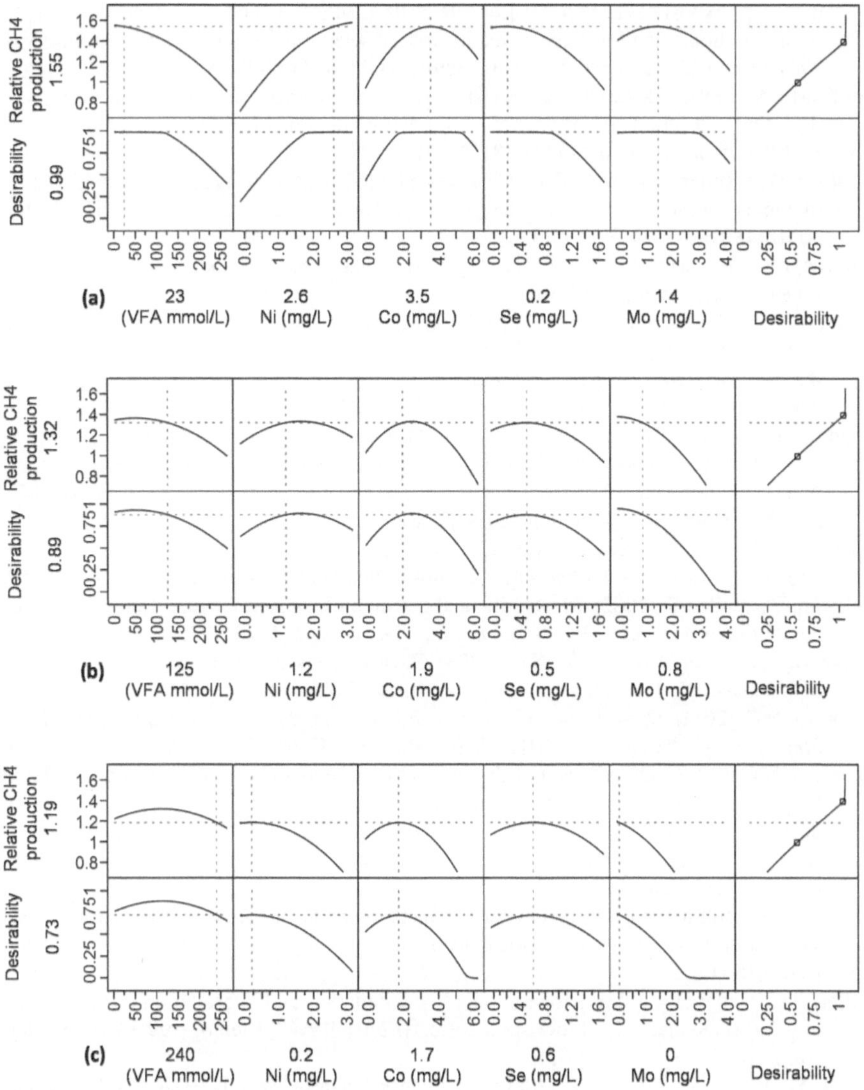

FIGURE 2.7 Optimum TEs configuration for CH_4 production at 37°C due to Ni, Co, Se and Mo supplementation to VFA levels (a) 23 mmol/L; (b) 125 mmol/L; and (c) 240 mmol/L.

REFERENCES

Amani, T., Nosrati, M., Mousavi, S.M., Kermanshahi, R.K. (2010). Study of syntrophic anaerobic digestion of volatile fatty acids using enriched cultures at mesophilic conditions. *International Journal of Environmental Science and Technology* 8, 83–96. Doi: 10.1007/BF03326198.

Appels, L., Baeyens, J., Degrève, J., Dewil, R. (2008). Principles and potential of the anaerobic digestion of waste-activated sludge. *Progress in Energy and Combustion Science* 34, 755–781. Doi: 10.1016/j.pecs.2008.06.002.

Ariunbaatar, J., Esposito, G., Yeh, D., Lens, P. (2016). Enhanced anaerobic digestion of food waste by supplementing trace elements: role of selenium (vi) and iron (ii). *Frontiers in Environmental Science* 4, 8. Doi 10.3389/fenvs.2016.00008 ISSN 2296665X.

Bardi, M., Aminirad, H. (2020). Synergistic effects of co-trace elements on anaerobic co-digestion of food waste and sewage sludge at high organic load. *Environmental Science and Pollution Research* 27 (15), 18129–18144.

Batstone, D.J., Keller, J., Newell, R.B., Newland, M. (2000). Modelling anaerobic degradation of complex wastewater. I: model development. *Bioresources Technology* 75, 67–74. Doi: 10.1016/S0960-8524(00)00018-3.

Burgess, J.E., Pletschke, B.I. (2008). Hydrolytic enzymes in sewage sludge treatment: a mini-review. *Water SA* 34, 343–350.

Caspi, R., Foerster, H., Fulcher, A., Kaipa, P., Krummenacker, M., Latendresse, M., Paley, S., Rhee, Y., Shearer, G., Tissier, C., Walk, C., Zhang, P., Karp, D. (2008). The MetaCyc database of metabolic pathways and enzymes and the bioCyc collection of pathway/genome databases. *Nucleic Acids Research* 36, 623–631.

Chan, P., Lu, Q., de Toledo, R., Gu, J., Shim, H. (2019). Improved anaerobic co-digestion of food waste and domestic wastewater by copper supplementation – Microbial community change and enhanced effluent quality. *Science of the Total Environment* 670, 337–344. Doi 10.1016/j.scitotenv.2019.03.081.

Chen, Y., Cheng, J.J., Creamer, K.S. (2008). Inhibition of anaerobic digestion process: a review. *Bioresource Technology* 99, 4044–4064. Doi: 10.1016/j.biortech.2007.01.057.

Colleran, E., Pender, S. (2002). Mesophilic and thermophilic anaerobic digestion of sulphate-containing wastewaters. *Water Science and Technology* 45, 231–235.

Drake, H.L., Gössner, A.S., Daniel, S.L. (2008). Old acetogens, new light. *Ann. N. Y. Academy of Science* 1125, 100–28. Doi: 10.1196/annals.1419.016.

Ezebuiro, N.C. (2014). *Optimization of Anaerobic Digestion: Influence of Trace Elements on Methanization Processes.* TUHH-GFEU. (Band 87). ISBN 978-3942768-122, Hamburg.

Ezebuiro, N.C. (2018). Dynamics and the interaction effects of trace elements and volatile fatty acids concentrations on methanization processes during thermophilic anaerobic digestion. *Nigerian Journal of Biotechnology* 35, 53–61. Doi: 10.4314/njb.v35i1.7.

Ezebuiro, N.C., Körner, I. (2017). Characterisation of anaerobic digestion substrates regarding trace elements and determination of the influence of trace elements on the hydrolysis and acidification phases during the methanisation of a maize silage-based feedstock. *Journal of Environmental Chemical Engineering* 5, 341–351. Doi: 10.1016/j.jece.2016.11.032.

Gallert, C., Winter, J. (2008). Propionic acid accumulation and degradation during restart of a full-scale anaerobic biowaste digester. *Bioresource Technology* 99, 170–178. Doi: 10.1016/j.biortech.2006.11.014.

Gnida, M., Ferner, R., Gremer, L., Meyer, O., Meyer-Klaucke, W. (2003). A novel binuclear [CuSMo] cluster at the active site of carbon monoxide dehydrogenase: characterization by x-ray absorption spectroscopy. *Biochemistry* 42, 222–230. Doi: 10.1021/bi026514n.

Gustavsson, J., Svensson, B.H., Karlsson, A. (2011). The feasibility of trace element supplementation for stable operation of wheat stillage-fed biogas tank reactors. *Water Science and Technology* 64, 320–325. Doi: 10.2166/wst.2011.633.

He, Z., Liu, W., Gao, Q., Tang, C., Wang, L., Guo Z., Zhou, A., Wang, A. (2018). Potassium ferrate addition as an alternative pre-treatment to enhance short-chain fatty acids production from waste activated sludge. *Bioresource Technology* 247, 174–181. Doi 10.1016/j.biortech.2017.09.073.

Heikinheimo, P., Tuominen, V., Ahonen, A., Teplyakov, A., Cooperman, B., Baykov, A., Lahti, R., Goldman, A. (2001). Toward a quantum-mechanical description of metal-assisted phosphoryl transfer in pyrophosphatase. *Proceedings of the National Academy of Sciences, U. S. A.* 98, 3121–3126. Doi: 10.1073/pnas.061612498.

Higuchi, Y., Ohashi, A, Imachi, H., Harada, H. (2005). Hydrolytic activity of alpha-amylase in anaerobic digested sludge. *Water Science and. Technology* 52 (1–2), 259–266.

Hullebusch, E.D., Utomo, S., Zandvoort, M.H., Piet, P.N. (2005). Comparison of three sequential extraction procedures to describe metal fractionation in anaerobic granular sludges. *Talanta* 65, 549–558. Doi: 10.1016/j.talanta.2004.07.024.

Ingram-Smith, C., Gorrell, A., Lawrence, S.H., Iyer, P., Smith, K., Ferry, J.G. (2005). Characterization of the acetate binding pocket in the Methanosarcinathermophila acetate kinase. *Journal of Bacteriology* 187, 2386–2394. Doi: 10.1128/JB.187.7.2386-2394.2005.

Iranpour, R., Alatriste-Mondragon, F., Cox, H.H.J., Haug, R.T. (2005). Effects of transient temperature increases on odor production from thermophilic anaerobic digestion. *Water Science and Technology* 52, 229–35.

Kleerebezem, R., Stams, A.J.M. (2000). Kinetics of syntrophic cultures: a theoretical treatise on butyrate fermentation. *Biotechnology and Bioengineering* 67, 529–543. Doi: 10.1002/(SICI)1097-0290(20000305)67:5<529::AID-BIT4>3.0.CO;2-Q.

Li, J., Ban, Q., Zhang, L., Jha, A.K. (2012). Syntrophic propionate degradation in anaerobic digestion: a review. *International Journal of Agriculture and Biology* 14, 843–850.

Lindahl, P. A. (2004). Acetyl-coenzyme A synthase: the case for a Ni2+-based mechanism of catalysis. *Journal of Biological Inorganic Chemistry* 9, 516–24. Doi: 10.1007/s00775-004-0564-x.

Magdalena, J., Greses, S., González-Fernández, C. (2019). Impact of organic loading rate in volatile fatty acids production and population dynamics using microalgae biomass as substrate. *Scientific Reports* 9 (1), 18374. Doi 10.1038/s41598-019-54914-4.

Mata-Alvarez, J. (2003). Fundamentals of the anaerobic digestion process, in: Mata-Alvarez, J. (Ed.), *Biomethanization of the Organic Fraction of Municipal Solid Waste.* IWA Publishing Company, Amsterdam, pp. 1–19.

Meegoda, J., Li, B., Patel, K., Wang, L. (2018). A review of the processes, parameters, and optimization of anaerobic digestion. *International Journal of Environmental Research and Public Health* 15(10), 2224. Doi: 10.3390/ijerph15102224.

Ortner, M., Leitzinger, K., Skupien, S., Bochmann, G., Fuchs, W. (2014). Efficient anaerobic mono-digestion of N-rich slaughterhouse waste: influence of ammonia, temperature and trace elements. *Bioresource Technology* 174, 222–232. Doi 10.1016/j.biortech.2014.10.023.

Pobeheim, H., Munk, B., Johansson, J., Guebitz, G.M. (2010). Influence of trace elements on methane formation from a synthetic model substrate for maize silage. *Bioresource Technology* 101, 836–839. Doi: 10.1016/j.biortech.2009.08.076.

Ragsdale, S.W., Pierce, E. (2008). Acetogenesis and the Wood-Ljungdahl pathway of CO_2 fixation. *Biochimica et Biophysica Acta* 1784, 1873–1898. Doi: 10.1016/j.bbapap.2008.08.012.

Rajagopal, R., Béline, F. (2011). Anaerobic hydrolysis and acidification of organic substrates: determination of anaerobic hydrolytic potential. *Bioresource Technology* 102, 5653–5658. Doi: 10.1016/j.biortech.2011.02.068.

Reda, T., Plugge, C.M., Abram, N.J., Hirst, J. (2008). Reversible interconversion of carbon dioxide and formate by an electroactive enzyme. *Proceedings of the National Academy of Sciences, U. S. A.* 105, 10654–10658. Doi: 10.1073/pnas.0801290105.

Schink, B., Stams, A.J.M. (2012). Syntrophism among prokaryotes. in: *The Prokaryotes: Prokaryotic Communities and Ecophysiology*, pp. 471–493. Doi: 10.1007/978-3-642-30123-0_59.

Scholten, J.C.M., Conrad, R. (2000). Energetics of syntrophic propionate oxidation in defined batch and chemostat co-cultures. *Applied Environmental Microbiology* 66, 2934–2942. Doi: 10.1128/AEM.66.7.2934-2942.2000.

Seravalli, J., Ragsdale, S.W. (2000). Channelling of carbon monoxide during anaerobic carbon dioxide fixation. *Biochemistry* 39, 1274–1277. Doi: 10.1021/bi991812e.

Shima, S., Warkentin, E., Thauer, R.K., Ermler, U. (2002). Structure and function of enzymes involved in the methanogenic pathway utilizing carbon dioxide and molecular hydrogen. *Journal of Biosciences and Bioengineering* 93, 519–530. Doi: 10.1016/S1389-1723(02)80232-8.

Siegrist, H., Renggli, D., Gujer, W. (1993). Mathematical modelling of anaerobic mesophilic sewage sludge treatment. Water Science and Technology 27 (2), 25–36. Doi: 10.2166/wst.1993.0070.

Uemura, S. (2010). Mineral requirements for mesophilic and thermophilic anaerobic digestion of organic solid waste. *International Journal of Environmental Research* 4, 33–40.

Voss, E., Weichgrebe, D., Rosenwinkel, K.H. (2009). *FOS/TAC-Deduction, Methods, application and Significance, in: InternationaleWissenschaftskonferenz "Biogas Science 2009 – Science Meets Practice"*. LfL-Bayern, Erding.

Wainaina, S., Lukitawesa, K.A.M., Taherzadeh, M. (2019). Bioengineering of anaerobic digestion for volatile fatty acids, hydrogen or methane production: a critical review. *Bioengineered* 10 (1), 437–458. Doi 10.1080/21655979.2019.1673937.

Yazdanpanah, A., Ghasimi, D., Kim, M., Nakhla, G., Hafez, H. (2018). Impact of trace element supplementation on mesophilic anaerobic digestion of food waste using Fe-rich inoculum. *Environmental Science and Pollution Research* 25 (29), 29240–29255.

Zandvoort, M.H., Geerts, R., Lettinga, G., Lens, P.N.L. (2003). Methanol degradation in granular sludge reactors at sub-optimal metal concentrations: role of iron, nickel and cobalt. *Enzyme and Microbial Technology* 33, 190–198. Doi: 10.1016/S0141-0229(03)00114-5.

Zhang, Y., Walker, M., Banks, C. (2010). Technical Report: Optimising Processes for the Stable Operation of Food Waste. Civ. Eng. Defra project code WR1208.

Ziemiński, K., Frac, M. (2012). Methane fermentation process as anaerobic digestion of biomass: transformations, stages and microorganisms. *African Journal of Biotechnology* 11, 4127–4139. Doi: 10.5897/AJBX11.054.

3 Appraising the Bioactivity Potentials of Eco-Friendly Synthesized Metallic Nanoparticles

Mike O. Ojemaye
University of Fort Hare

Anthony I. Okoh
University of Sharjah
University of Fort Hare

CONTENTS

DOI: 10.1201/9781003180289-3

3.1 INTRODUCTION

Nanotechnology is the science and engineering of the design, synthesis, characterization and application of small sized materials and devices with dimensions in the nanometer range, and because of their sizes nanoparticles have unique biological and physicochemical properties. They exhibit better properties than their bulk materials (Ojemaye et al., 2017; Varadavenkatesan et al., 2017). It is well recognized that chemistry procedures are not environmentally friendly because they entail synthesis that employs huge amounts of organic solvents and materials that must be mined from the earth among others, which have taken a huge toll on environmental and public health at large. Over the years, there have been greater calls for the use of greener and environmentally friendly processes, and this is likely to continue to grow in the days ahead. Environmentally friendly synthesis of metallic nanoparticles involves the design of metallic nanoparticles with or without reduced use of hazardous substances or chemicals. This synthesis option is also known as green synthesis according to US EPA (2020) and helps to prevent pollution at molecular level by reducing the adverse impacts of chemical processes and products on the environment and human health. Green Chemistry is based on 12 principles with less hazardous synthesis of materials taking a prominent role in number 3: to design synthesis and produce materials/compounds with little or no side effects to the environment or man.

Metallic nanoparticles possess varying physicochemical properties (including thermal, surface, electrical and optical properties) which are different from their original bulk compounds in many aspects. They are usually synthesized by the reduction and/or precipitation of their metal salts or by physical means, and their size, shape, composition, synthesis route, surface properties, stability and purity are some factors that determine their biological activities (Teske and Detweiler, 2015; Wang et al., 2016). However, majority of the chemicals employed for their reduction/precipitation are not environmentally friendly as they are flammable, toxic, have poor yield and are known to have disposal problems. Examples of metallic nanoparticles are silver nanoparticles (AgNPs), gold nanoparticles (AuNPs), copper nanoparticles (CuNPs), zinc oxide nanoparticles (ZnONPs), platinum nanoparticles and titanium dioxide nanoparticles (TiO_2 NPs). Due to the increasing importance of metallic nanoparticles fabricated through the green synthesis methods and the attendant challenges with chemical methods of synthesis of metallic nanoparticles, there have been increasing attention towards new strategies for the synthesis of these classes of nanoparticles including green synthesis method (Elegbede and Lateef, 2019). This chapter systematically highlights the biological importance of metallic nanoparticles as well as the factors that enhance their biological activities. Prior to these, discussions on the mechanism of their synthesis are reported as well as the limitations and advantages of biosynthesized metallic nanoparticles. Lastly, the present state of green and sustainable chemistry in Africa and the need for green chemistry policy in Africa conclude this chapter.

3.2 GREEN SYNTHESIS OF METALLIC NANOPARTICLES

Few synthesis methods of metallic nanoparticles including the reduction of metal salts with established reducing agents like borohydrides, citrates, alcohols, alkyl sulfates, etc. are already known. These metallic nanoparticles obtained from chemical route exhibit some aggregation and so compromise the size factor after a period of time (Mukherjee et al., 2008). Furthermore, this method of synthesis is not environmentally friendly as it involves the use of hazardous chemicals such as hydrazine, which is cytotoxic and carcinogenic, as well as very severe synthesis conditions including high temperatures. In addition, the chemicals employed in this method are very expensive and the method itself requires high energy. On the other hand, physical method requires high temperature, high pressure and expensive equipment which make it not a scalable and profitable method. Therefore, green synthesis method is an alternate, safe and convenient method of synthesis of these types of nanoparticles as it involves the use of non-hazardous solvents at room temperature and biological pH. They can be controlled and manipulated to suit the application for which they were designed. Several green synthesis routes are known which are based on the substrate that induces the reduction of the metal salts and stabilization of the metallic nanoparticles (for example, plant parts and microorganisms). This serves as the main principle of green synthesis of metallic nanoparticles. In this section, the different types of nanoparticles synthesis via green route are discussed.

3.2.1 Microorganisms-Based Synthesis

The synthesis of metallic nanoparticles using microorganisms is one of the green synthesis methods that have generated a lot of attention recently (Yadi et al., 2018). However, reduction of metal salts with the aid of microorganisms usually yields metallic nanoparticles via three main routes: (1) using bacteria and unicellular organisms in which metallic nanoparticles are deposited in the periplasmic and intracellular space, which are then treated with detergents and/or ultrasonication to produce pure nanoparticles; (2) when filamentous fungi are employed, the challenges observed with the use of bacteria and unicellular organisms are avoided and hence have been preferred to the former; nevertheless, the use of fungi has its own limitation as the nanoparticles generated by this route are toxic to humans and difficult to handle (Casadevall and Pirofski, 1999; Mukherjee et al., 2008); and (3) the use of algae where about 1,500 species have been identified to be very useful for metallic nanoparticles synthesis (Yurkov et al., 2011).

3.2.1.1 Bacteria Biosynthesized Metallic Nanoparticles

The use of bacteria for the synthesis of metallic nanoparticles has been extensively explored. They have huge ability to reduce metal ions and stabilize their corresponding metallic nanoparticles. For this purpose, a wide array of bacterial species has been employed with the prokaryotic and actinomycetes group mostly reported

(Singh et al., 2018). The cell wall of bacteria plays a huge role here as metals are expected to penetrate the wall into the cytoplasm and be taken back for extracellular release via the wall meshwork. The interaction between bacterial cell and metal ion is enhanced by the fact that peptidoglycan in the cell wall of bacteria is very negatively charged and bonds electrostatically with the positively charged metal ion. Although the alteration of the wall of the bacteria is possible by treatment with anhydride in order to target carboxyl and amine groups, this alteration makes the change from positive charge to negative charge occur in the cell wall of the organism. An example of this alteration is the one carried out on peptidoglycan of *B. subtilis* in which the carboxyl group present in its glutamic acid is converted for metallic interaction (Beveridge and Murray, 1980). The ease and possibility of manipulation/alteration of bacteria makes them advantageous for the synthesis of metallic nanoparticles via the green route. The exact morphology of bacteria and cell intracellular and extracellular environment are the major factors that determine the phases of the synthesized metallic nanoparticles (i.e. crystalline and non-crystalline) (Das et al., 2017). A number of products of nanosized materials biosynthesized from bacteria are now being used in agriculture as nano-fertilizers, nano-pesticides and nano-coatings. The main challenges encountered in the agricultural sector are the excessive use of chemical fertilizers, decrease of soil fertility as well as availability of water. The use of nano-fertilizers ensures an increase in soil nutrients and decreases soil toxicity and the general effects encountered with fertilizers obtained from chemicals (Salem and Fouda, 2021).

3.2.1.2 Fungi Biosynthesized Metallic Nanoparticles

An important method for the preparation of well-dispersed metallic nanoparticles is the fungi fabricated synthesis route. The synthesis of metallic nanoparticles using fungi, *Fusarium oxysporum*, may be intracellular or extracellular. For extracellular synthesis of metallic nanoparticles using fungi, the metal salt will be added to water solution containing only the fungal biomolecules, and thereafter free metallic nanoparticles are formed in the dispersion. The nanoparticles dispersion will then be purified (using filtration, centrifugation, dialysis and many more) to enable the elimination of impurities and fungal residues. This method is very direct and widely used because it averts the difficulty encountered with the release of the nanoparticles to the cells (Azmath et al., 2016; Sabri et al., 2016; Costa Silva et al., 2017; Gudikandula et al., 2017). In intracellular synthesis of metallic nanoparticles, extraction of the synthesized metallic nanoparticles is carried out by treatment with chemicals, and then centrifugation and filtration to separate the nanoparticles from the biomass. Here, the nanoparticles are synthesized by adding metal salt into mycelial culture followed by internalization in the biomass (Castro-Longoria et al., 2011; Molnár et al., 2018). Fungi are advantageous than bacteria in the synthesis of metallic nanoparticles because of the biological agents present in them and the large amount of intracellular enzymes, proteins and reducing substances on their cell walls (Singh et al., 2018). This is possible because of the combination of two of their properties: (1) enzymes inside the fungal cell and in their cell wall provide the advantage of wall binding ability, high metal tolerance and intracellular metal adsorption capacity (Dias et al., 2002); and (2) their ability to withstand pressure, flow and agitation in reaction chambers or bioreactors in contrast to bacteria and plants. AuNPs obtained from fungi are currently being employed as drug delivery systems.

3.2.1.3 Algae Biosynthesized Metallic Nanoparticles

Algae are predominant eukaryotic primary producers in marine environments. They produce their food via photosynthesis generating oxygen as by-products. Some algae, specifically *Chlorella* sp., have been observed to harbour an array of heavy metals, and studies on the adsorption of heavy metals suggest that brown algae are more efficient than other algae and autotrophs (Mata et al., 2008). The ability of algae to accumulate heavy metals makes them important candidates for metal reduction, a process that is very vital for metallic nanoparticles synthesis, through either the green or chemical route. For instance, *Chlorella vulgaris*, a green algae with single cell and strong anti-tumour properties, was found to exhibit powerful binding capability towards chloride (III) salt of gold. Because of this, an algal gold composite was formed which allowed for the reduction of Au(III) to Au(0) (Velusamy et al., 2016). Interestingly, almost all algal gold composites result in metallic AuNPs with clear crystals of gold obtained in the exterior and interior surfaces. Recently, *Tetraselmis kochinensis* was used to produce AuNPs intracellularly (Das et al., 2017), while Mata et al. (2009) also reported the extracellular reduction of gold from +3 to 0 with the aid of the biomass of brown alga, *Fucus vesiculosus*. Worthy of note is that several metallic nanoparticles such as CuNPs, ZnO and Fe_2O_3, which have been vastly obtained from bacteria and plant parts, have not been successfully synthesized using algae (Sau and Murphy, 2004). Protein molecules in the cell of *Oscillatoria willei*, a cyanobacteria, were reported to have been responsible for reducing Ag ion to metallic Ag (Ali et al., 2011). Au nanoparticles synthesized using algae have been employed for X-ray imaging primarily as X-ray retention coefficient, nontoxicity, simplicity of engineered control, surface functionalization for colloidal dependability and focused on conveyance (Gour and Jain, 2019). A summary of the variety of metallic nanoparticles synthesized using microorganisms is presented in Table 3.1.

3.2.2 Plant-Based Synthesis

The synthesis of metallic nanoparticles using the extract of plant is currently being exploited for the generation of environmentally benign nanoparticles. The plant extracts may function as reducing as well as capping agents during this synthesis. The different water-soluble metabolites and biomolecules in plants have been identified as responsible for the reduction of metal ions, and these include terpenoids, phenolic compounds, alkaloids, amino acids, proteins, enzymes, polysaccharides, etc. This synthesis method is very economical as it is very cheap and could be easily scaled up for production of metallic nanoparticles for industrial use. Extracts of a variety of different species of plants and their parts have been successfully employed for metallic nanoparticles synthesis in our previous reports (Okaiyeto et al., 2019; Rotimi et al., 2019; Ojemaye et al., 2021). One advantage in the use of plants for the synthesis of metallic nanoparticles is that any of the plant parts (leaf, stem bark, root, seed, fruit, etc.) can be used for this purpose, and they have all proven to be very effective

TABLE 3.1
Some Metallic Nanoparticles Synthesized Using Microorganisms

Organism/ Species	Metallic NPs	Morphology	Size (nm)	Application	References
			Bacteria		
Pseudomonas proteolytica	Ag	Spherical	5–12	Antibacterial activity against *E. coli, Arthrobacter kerguelensis, Arthrobacter gangotriensis, Pseudomonas antarctica*	Shivaji et al. (2011)
Aeromonas hydrophila	ZnO	Spherical	57.7	Antibacterial activity	Jayaseelan et al. (2012)
Pseudomonas aeruginosa	ZnO	Spherical	35–80	Antioxidant activity	Singh et al. (2014)
Bacillus licheniformis MTCC9555	ZnO	Flower	250	-	Tripathi et al. (2014)
			Fungi		
Aspergillus terreus	ZnO	Spherical	54.8–82.6	Antifungal activity	Baskar et al. (2013)
Aspergillus fumigatus JCF	ZnO	Spherical	60–80	Antibacterial activity	Mohd Yusof et al. (2019)
White rot fungi	Ag	Colloidal	15–25	Antibacterial activity	Gudikandula et al. (2017)
Filamentous fungi	Au	-	< 40	Antibacterial activity	Molnár et al. (2018)
Algae					
Neochloris oleoabundans	Ag	Quasi-spherical	16.63	Antibacterial activity	Lan and Bao (2019)
Scenedesmus abundans	Ag	-	59–66	Antibacterial activity	Aziz et al. (2014)
Padina pavonica	Ag	Spherical	54	Antibacterial activity	Sahayaraja and Rajesh (2012)

for nanoparticle formation. These different plant parts are known to be rich in biomolecules needed for metal ion reduction and capping. For instance, the seed extract of *Syzygium cumini* plant was used for the synthesis of AgNPs (Banerjee, 2011). The seed extract was found to have good antioxidant activity, but the nanoparticles obtained through the seed extract were observed to have greater antioxidant activity than the seed extract, probably due to preferential attachment of the phytochemicals in the plant extract (responsible for its antioxidant property) to the surface of the nanoparticles (Banerjee, 2011). *Dioscorea bulbifera* tuber extract was employed for AuNPs synthesis of various morphologies. The synthesized nanoparticles were

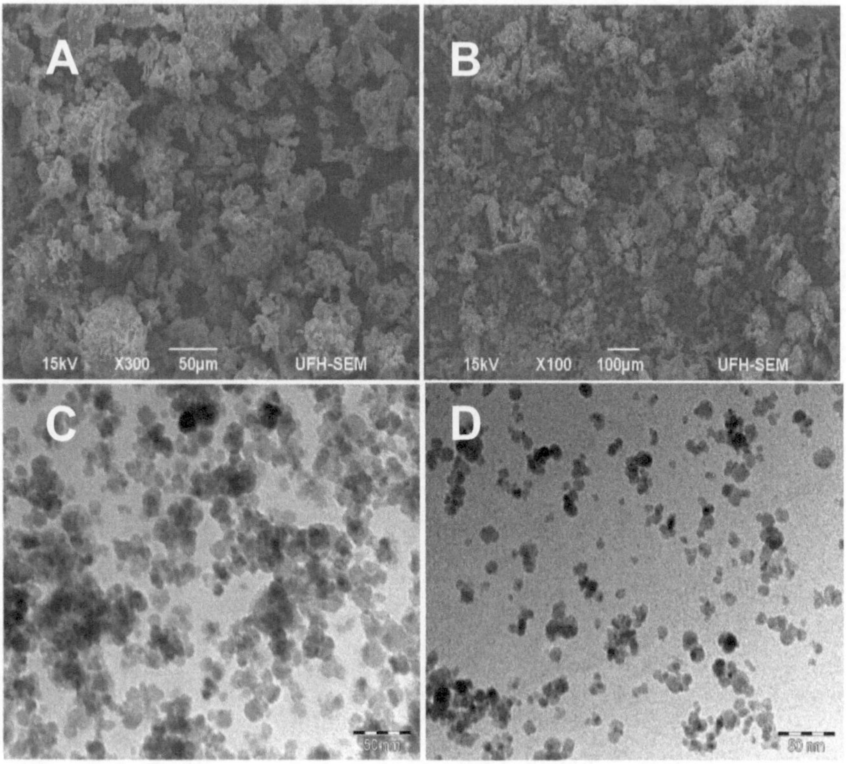

FIGURE 3.1 CuONPs mediated by the plant parts of *Crataegus ambigua*. SEM images of (a) fruit CuONPs (b) leaf CuONPs and TEM images of (c) fruit CuONPs (d) leaf CuONPs.

combined with known antibiotics and found to exhibit synergistic activity against *Escherichia coli*, *Pseudomonas aeruginosa* and *Acinetobacter baumannii* (Ghosh et al., 2012). Also, synthesis of AgNPs with the leaf extract of *Piper longum* was reported by Jacob et al. (2012). The nanoparticles exhibited huge cytotoxic effect on Hela cancer cells. Worthy of mention is that the plant parts do not influence the surface morphology of the obtained nanoparticles. We recently reported the synthesis of AuNPs and AuNPs using the extracts of leaf, seed and flower of *Callistemon citrinus* (Larayetan et al., 2019; Rotimi et al., 2019). Our study revealed that the synthesized nanoparticles were spherical in shape with little agglomeration. This type of nanoparticles obtained from plants has had application in Dentistry. For instance, AgNPs are presently used in dental instruments and swathes (Gour and Jain, 2019). Figure 3.1 shows the scanning electron microscope (SEM) and transmission electron microscope (TEM) images of copper oxide nanoparticles (CuONPs) synthesized from the extract of leaf and fruit parts of *Crataegus ambigua* plant. From this figure, we can observe that the morphology of these nanoparticles was similar and that the different plant parts did not in any way influence the morphology of the nanoparticles. Table 3.2 shows the different plants reported for the synthesis of metallic nanoparticles.

TABLE 3.2
Biosynthesized Metallic Nanoparticles Mediated by Plant

Plants	Metallic NPs	1. Morphology	2. Size (nm)	3. Application	4. References
Flower extract of *Nyctanthes arbor-tristis*	ZnO	Spherical	12–32	Antifungal activity against *Alternaria alternata* (ITCC 6531), *Aspergillus niger* (ITCC 7122), *Botrytis cinerea* (ITCC 6192), *Fusarium oxysporum* (ITCC 55) and *Penicillium expansum* (ITCC 6755)	Jamdagni et al. (2018)
Deverra tortuosa	ZnO	Spherical	31.18	Anticancer activity against two cell lines (human colon adenocarcinoma “Caco-2” and human lung adenocarcinoma “A549”)	Selim et al. (2020)
Leaf extract of *Salvia officinalis*	Ag	Spherical	41	Antiplasmodial activity against *Plasmodium falciparum* (strain 3D7)	Okaiyeto et al., (2020)
Leaf, seed and flower parts of *Calistemon citrinus*	Au/Ag	Spherical	35.5 Au/29 Ag	Antibacterial, antiplasmodial and antitrypanocidal activities	Larayetan et al. (2019), Rotimi et al. (2019)
Leaf part of *Oedera genistifolia*	Ag/ AgCl	Spherical	34.2	Antibacterial activity against some gram negative and positive bacteria	Okaiyeto et al. (2019)
Flower part of *Millettia pinnata* plant	Cu	Spherical	23 ± 2	Antibacterial, antioxidant and anti-inflammatory activities using albumin denaturation and membrane stabilization	Thiruvengadam et al. (2019)
Seed part of *Psoralea corylifolia*	Fe_2O_3	Spherical and rod-like shape	< 39	Anticancer activity	Nagajyothi et al. (2017)
Thymus kotschyanus	Ag	Crystalline	50–60	Antibacterial and antioxidant activities	Hamelian et al. (2018)
Leaf extract of *Achillea biebersteinii*	Cu	Spherical	16.8	Antioxidant and neurological potentials against methamphetamine-induced cell death in PC12	Wang et al. (2020)
Scutellaria barbata	Au	Spherical	25	Anticancer activity	Wang et al. (2019)
Polygonium minus	Cu	-	20–30	Antibacterial activity	Ullah et al. (2019)
Terminalia catappa	Ag	Spherical	12–40	Antioxidant and anti-inflammable activities	Mohamed El-Rafie and Abdel-Aziz (2014)

3.3 MECHANISMS OF GREEN SYNTHESIS OF METALLIC NANOPARTICLES

The mechanism of formation of metallic nanoparticles differs from organism to organism and to plants. However, the formation of metallic nanoparticles follows similar route irrespective of the organism or plant involved. For instance, synthesis of nanoparticles using the extract of organisms involves the mixing of the extract of the organisms (intracellular or extracellular) with metal salt solution at ambient temperature. By doing this, the reduction of the metal ions takes place (Yin et al., 2016). The negatively charged carboxylic group inherent in enzymes on the organism's cell wall causes an attraction with the positively charged metal ions by means of electrostatic force, which now allows the enzyme to reduce the metal ions to its zero charged form; this zero charged form later matures and grows by accumulation and reduction (Ali et al., 2019).

Figure 3.2 describes the mechanism of synthesis of metallic nanoparticles. Here, the metal precursor is bio-reduced by enzymes from microorganism or carbohydrates and phytochemicals from plants. The reduction process converts the metal ion to elemental metal. This elemental metal then matures to nanoparticles. However, specific mechanisms are observed for different substrates of metallic nanoparticles, and these are described below.

3.3.1 Microorganism Extract Mechanism

For nanoparticles synthesis using microorganisms, a number of mechanisms can come into play. The first is the capture of the metal ion inside or on the surface

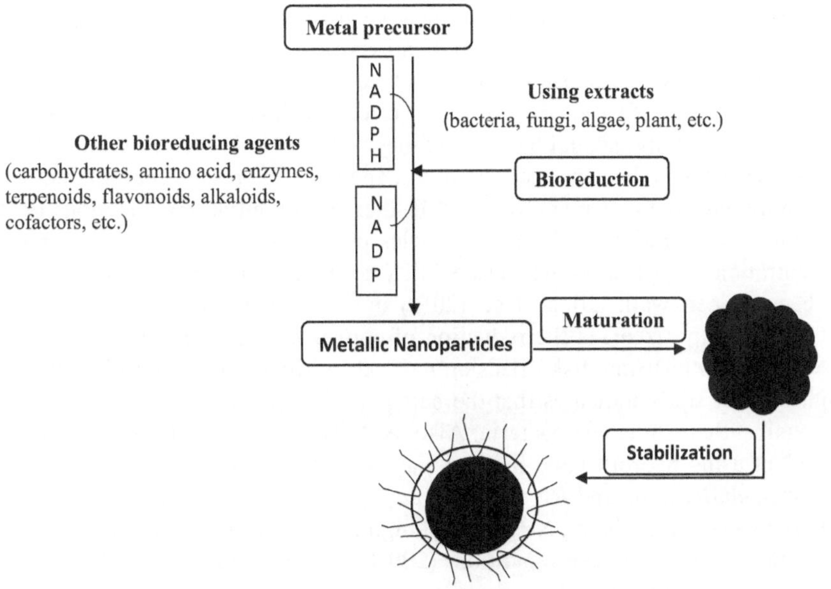

FIGURE 3.2 General mechanism for biosynthesis of metallic nanoparticles.

of the cell of the microorganism followed by the reduction of the metal ions by enzymes present in the cell. Velusamy et al. (2016) reported the mechanism for the synthesis of metallic nanoparticles using microorganisms as having the metal ion captured by the inside of the microbial cell through electrostatic interaction to be acted upon by the cell enzymes to cause a reduction of the metal ion to its nuclei; this will then grow into metallic nanoparticles. Be that as it may, this bio-reduction mechanism involving microorganisms remains largely under-utilized for the synthesis of nanoparticles.

3.3.2 Plant Extract Mechanism

For metallic nanoparticles synthesis using plant extract, so many factors determine its success (i.e. rate of formation of nanoparticles, properties of the nanoparticles, stability and yield). These factors include phytochemicals type, concentration of phytochemicals, pH, concentration of metal salt and reaction temperature. The mechanism of synthesis involves the mixing of plant extract with the solution of metal salt at controlled reaction conditions. This synthesis method is much shorter to complete compared to microorganism-based approach, which requires longer completion time (Dwivedi and Gopal, 2010).

3.4 BIOLOGICAL IMPORTANCE OF ENVIRONMENTALLY FRIENDLY SYNTHESIZED METALLIC NANOPARTICLES

3.4.1 Antibacterial Importance of Biosynthesized Metallic Nanoparticles

Metallic nanoparticles from biogenic means have been widely employed as antibacterial agents among the many applications of these classes of nanoparticles because they have the capacity to disrupt cell membrane's polymer subunit in harmful organisms. One of the multiplier effects of this is that the nanoparticles break the cell membrane and halt the synthesis of protein in the bacterial cell. The interaction between bacteria and metallic nanoparticles is such that the nanoparticles bind with the active site of the cell, thereby inhibiting the functions of the cell. The concentration of the nanoparticles plays a major role in this as higher concentration allows for faster permeability of the cell membrane and so breaks the bacteria cell wall. Ullah et al. (2019) investigated the antibacterial activity of CuNPs synthesized via the reduction of copper nitrate by the plant extract of *Polygonum minus* using disk diffusion method for three different organisms. The result of their study indicates that the nanoparticles showed large zone of inhibition against their three test bacteria. Also, AgNPs prepared via the extract of aloe vera plant using hydrothermal method was evaluated for its antibacterial activity against *S. epidermidis* and *Ps. aeruginosa* (Tippayawat et al., 2016). Although the authors reported that the AgNPs showed high antibacterial activity, they ascribed this to the varying synthesis conditions (200°C for 720 minutes and 100°C for 360 minutes) adopted in the study.

3.4.2 Antiplasmodial Importance of Biosynthesized Metallic Nanoparticles

Protozoan parasites (genus: *Plasmodium*) are usually transmitted when infected Anopheles mosquitoes bite humans and cause malaria. There are different species of *Plasmodium* that cause disease and include *Plasmodium vivax*, which is widespread, and *Plasmodium falciparium*, which causes most deaths. Other known *Plasmodium* species include *Plasmodium ovale*, *Plasmodium malariae*, *Plasmodium cynomolgi* and *Plasmodium knowlesi*. In Sub-Saharan Africa, malaria is a big health challenge, which has had dire consequences on most economies. In spite of huge global efforts to combat malaria through improved control interventions and enhanced funding for research on malaria, malaria deaths have continued to increase (Kojom et al., 2019). The development of resistance against artemisinin loci by *P. falciparum*, which was also encountered with previously used antimalarial drugs like sulfadoxine-pyrimethamine and chloroquine, has been a major setback to malaria prevention and elimination efforts globally. One of the diverse strategies for the development of new antimalarial drugs is the synthesis of metallic nanoparticles from plants and microorganisms. Reports from our group have shown that these green synthesized metallic nanoparticles, e.g. AgNPs and AuNPs, are very potent against *P. falciparum* parasite (Larayetan et al., 2019; Okaiyeto et al., 2019, 2020).

3.4.3 Anticancer Activity of Biosynthesized Metallic Nanoparticles

AuNPs biosynthesized from *Scutellaria barbata* were evaluated for their anticancer activity against pancreatic cancer cells (PANC-1). The AuNPs were reported to possess effective anticancer activity against PANC-1 (Wang et al., 2019). Biosynthesized AuNPs and gold-prodigiosin nanoparticles were reported by Dozie-Nwachukwu et al. (2017) for cancer treatment. The authors reported that the presence of prodigiosin, which is a secondary metabolite from *Serratia marcescens* and known to have good anticancer activity, enhanced the anticancer activity of the AuNPs. Although cancer is an un-inhibited spreading of tumour cells, the overexpression of the growth of the cell can be stopped using biosynthesized metallic nanoparticles.

3.4.4 Anti-inflammatory Importance of Biosynthesized Metallic Nanoparticles

Inflammation is a physical condition whereby an inflamed body part develops redness, sweating and pains in response to injury or infection (Agarwal et al., 2019). Tissue damages, wounds and infections require inflammatory response for them to heal. Two major parts of the defence mechanisms of the host mediate the inflammatory response: adaptive and innate immune response. The former is characterized by selectivity and aids the generation of immunological memory and removal of pathogens in the later phase, while the latter is the response to any foreign material by the primary host which is acted upon by other cells (Akira et al., 2006). Anti-inflammatory drugs are a class of compounds that prevent or aid to reduce clotting of blood. This is achieved by blocking cyclooxygenases (COX) involved in prostaglandins synthesis. Metallic nanoparticles have better persistence and effectiveness in treating inflammation since they penetrate inflammatory cells better. Biosynthesized metallic nanoparticles have better specificity to target sites including inflammatory tissues and cells (Viscido et al., 2014; Agarwal et al., 2019). They enter the cell via the cell membrane pores or ion channels. This however depends on the size of the metallic nanoparticles and occurs without the receptors, but because of electrostatic interactions, it involves adhesive interactions. For instance, David et al. (2014) reported the anti-inflammatory activity of AgNPs obtained from the bio-reduction of silver nitrate using the fruit extract of European black elderberry (*Sambucus nigra*). The anti-inflammatory activity of the biosynthesized AgNPs assessed in vitro on HaCaT cells and in vivo on acute inflammation model indicates that the AgNPs were effective as the cytokines production decreases in in vitro study. Similarly, the in vivo study showed the reduction of the cytokines and oedema levels in the paw tissues upon the pre-administration of AgNPs (David et al., 2014).

3.4.5 Antidiabetic Efficacy of Biosynthesized Metallic Nanoparticles

When a human has uncontrolled blood sugar level, the person is said to be suffering from dysfunctional metabolic condition known as diabetes mellitus. The treatment of this condition is still a challenge, but some foods or synthesized insulin drugs can be used to control the diabetes at a particular level. AuNPs among the metallic nanoparticles have shown to significantly reduce the level of liver enzymes like alkaline phosphate, alanine transaminase, etc. in mice observed to have diabetes (Saipriya and Daisy, 2012), suggesting that they could serve as alternate drugs for the treatment of diabetes mellitus. The diabetes model subjected to treatment using AuNPs showed a reduction in the level in glycosylated haemoglobin meant to sustain the normal range. The inhibition of acarbose sugar and α-amylase in animal model induced with diabetes was carried out with biosynthesized AgNPs obtained from the extract of *Sphaeranthus amaranthoides* (Swarnalatha et al., 2012). The study showed that the nanoparticles were able to control acarbose sugar in the animal model.

3.4.6 Antioxidant Activity of Green Synthesized Metallic Nanoparticles

The antioxidant potential of any material defends against the negative effect of reactive oxygen species (ROS). The ROS causes oxidative stress by cell damage when subjected to environmental stress conditions. They contain oxygen that are chemically extremely reactive (Hayyan et al., 2016). Usually, during the natural metabolism of oxygen, they are given off as by-products. The metallic nanoparticles from green route tend to nullify the adverse effect the ROS causes on living cells. Green synthesized AgNPs from *Vigna mungo* were examined for their antioxidant potentials. The study indicated that AgNPs showed free radical scavenging potential compared to their ascorbic acid standard (Varadavenkatesan et al., 2017).

3.4.7 Antitrypanocidal Activity of Green Synthesized Metallic Nanoparticles

Human African trypanosomiasis is known as African sleeping sickness, while the animal African trypanosomiasis in cattle is regarded as nagana. The latter is caused by a parasite known as *Trypanosoma brucei* and transmitted by *Tsetse* fly. The adverse effect of trypanosomosis on human health and economy is huge and requires continuous research on ways to curb this disease. Sadly, high production cost of chemical drugs, with the emergence of resistance strains and high incidence of their side effects, has rendered existing chemotherapy inefficient (Adelodun et al., 2013). Therefore, there is need to design other drugs especially of plant or microorganism origin. Few studies have reported the application of biosynthesized metallic nanoparticles against *Trypanosoma brucei*, including one from our group on the use of AuNPs synthesized from plant parts of *Callistemon citrinus* (Rotimi et al., 2019).

3.5 CHEMICAL REGULATION AND GREEN CHEMISTRY POLICY IN AFRICA

Synthetic chemicals that are being created together with natural inorganic and organic compounds are intentionally or accidentally released into the environment with humans and organisms on the receiving side of their harmful effects (Collins et al., 2020). There have been calls for the regulation and management of these chemicals taking into account the underlying principle of another field, specifically medicine, where practitioners agreed that "Good health can be achieved without doing any harm". This principle could be achieved in the field of science and chemistry in particular with green and sustainable chemistry. Green chemistry as a field is a small but important part of the overall effort in achieving an environmentally sustainable future, which is an intense goal that requires several disciplines and perspectives.

This field dates back to five decades, even though there were efforts to achieve similar goals in the field of chemistry since the early 1960s. This field started properly in 1990 as an aftermath of the United States Environmental Protection Agency (US EPA) Pollution Prevention Act titled: Alternative Synthetic Pathways for Pollution Prevention (US EPA, 2020). In 1998, in the United States, the 12 Principles of Green Chemistry were published, and this gave rise to conscious efforts to safe laboratory practices and provided frameworks for green chemistry forthwith (Cannon and Warner, 2011). Similarly, the United Nations General Assembly on 25 September 2015 approved the 2030 Agenda for Sustainable Development implementable under the 17 Sustainable Development Goals (SDGs). For instance, the task of the United Nations Industrial Development Organization (UNIDO) is to achieve environmental sustainability and shared prosperity (particularly connected to SDG 9) through sustainable and inclusive industrial development. However, the SDGs are universal set of goals and targets that United Nations Member Countries (which most African countries are members) are supposed to utilize when developing their own policies and agendas for the next 10 years (UNIDO, 2018; Schwager et al., 2016). While other advanced nations have long adopted/formulated green chemistry frameworks or policies in line with UNIDO's mandate, this cannot be said of Africa where green chemistry policy is not in existence, and where practiced, it is in its infancy. From literature, it is evident that a huge number of studies have been conducted on green chemistry synthesis (Elegbede and Lateef, 2019), but no unified platform existed until 2007 when green chemistry workshop was first held in Johannesburg, South Africa (SACI, 2014). This was closely followed by a congress held in Ethiopia in 2010 titled "the first Pan Africa Chemistry Network (PACN) Green Chemistry Congress" where the 12 principles of green chemistry were adopted to suit Africa's unique development. In 2011, the South African Chemical Institute (SACI) organized another conference at the University of Witwatersrand, Johannesburg, South Africa, in which the last day of the conference was exclusively about green chemistry (Jumbam and Jumbam, 2015). Despite the modest development in this aspect of chemistry in the continent, African countries are still some distance away from reaping the possible benefits of this field since there are no policies in place on green chemistry. To achieve this, deliberate policies need to be put in place on green chemistry in line with UNIDO's mandate by heads of government in Africa. In addition, it is important for African nations to begin to raise awareness and build capacity on the principles of green chemistry as well as promulgate laws to regulate hazardous chemical usage and handling. By so doing, patronization of green chemistry research will grow while also closing the gap between science-based research and real-world utilization of green and sustainable chemistry.

3.6 PROGRESS, DRAWBACKS AND FUTURE RECOMMENDATION

3.6.1 Progress

One of the major opportunities, which the green synthesis of metallic nanoparticles has brought, is the opening of natural nanomaterials factories. Two different aspects that require consideration in this process include: (1) selection of the most active microorganism/plant in terms of enzyme/biochemical activities with very robust enzyme systems and activities and higher chances to modulate efficient reduction of the captured metal ion in or on the surface of the cell to elemental metal needed for growth into nanoparticles; and (2) selection of the best conditions for enzyme activity, reaction and cell growth. Varying growth conditions such as light, nutrients, temperature, inoculum size, pH, buffer strength and mixing speed are very important to ensure effective nanoparticle synthesis in industrial scale (Iravani, 2011). The progress with green synthesized nanoparticles is the comparative advantage they possess over chemically synthesized metallic nanoparticles with respect to the former's broad application range including gene therapy, antibacterial and anticancer agents, magnetic resonance imaging, DNA analysis, targeted drug delivery systems and so on. These ranges of applications are not obtainable with chemically synthesized metallic nanoparticles because of the input of hazardous chemicals during their preparation (Li et al., 2011).

3.6.2 Drawbacks and Future Recommendation

The main setback in environmentally synthesized nanoparticles is the time/length of synthesis as it takes a while to grow the microbes employed and the synthesis of the nanoparticles and thus requires further research. In addition, the toxicity of these materials needs to be understood as many have been reported to be toxic while a few have been reported to be non-toxic. Emphasis should be on the study of the toxicity of the materials synthesized through this route so as to accrue real information for policy formulation and regulatory processes. Also, the synthesis of metallic nanoparticles of a specific shape and size through green synthesis route is challenging considering the fact that more optimization studies are required. Another drawback of synthesis route of metallic nanoparticles is that scaling-up production for commercialization is still at its infancy and in most places it is not even in existence which therefore discourages future research in this field. Finally, this review has suggested that green chemistry principles and knowledge can be successfully applied for metallic nanoparticles synthesis for applications to numerous biotechnological/biological opportunities, and this will enhance efforts towards green and sustainable world.

ACKNOWLEDGEMENT

The authors wish to appreciate South African Medical Research Council for research facilities.

REFERENCES

Adelodun, V. O. et al. (2013) Evaluation of antitrypanosomal and anti inflammatory activities of selected Nigerian medicinal plants in mice. *African Journal of Traditional, Complementary, and Alternative Medicines : AJTCAM / African Networks on Ethnomedicines*, 10(6): 469–476. Doi: 10.4314/ajtcam.v10i6.13.

Agarwal, H., Nakara, A. and Shanmugam, V. K. (2019) Biomedicine & Pharmacotherapy Anti-inflammatory mechanism of various metal and metal oxide nanoparticles synthesized using plant extracts : a review. *Biomedicine & Pharmacotherapy* 109: 2561–2572. Doi: 10.1016/j.biopha.2018.11.116.

Akira, S., Uematsu, S. and Takeuchi, O. (2006) Pathogen recognition. *Cell* 3: 783–801. Doi: 10.1016/j.cell.2006.02.015.

Ali, D. M., Sasikala, M., Gunasekaran, M. and Thajuddin, N. (2011) Biosynthesis and characterization of silver nanoparticles using marine cyanobacterium, oscillatoria willei ntdm01 Type Sec. amides, freeNH (trans) Charged amines (C=NH +) organo-silicon compounds aliphatic esters secondary alcohols vibration mode. *Digest Journal of Nanomaterials and Biostructures* 6(2): 385–390.

Ali, J., Ali, N., Wang, L., Waseem, H. and Pan, G. (2019) Revisiting the mechanistic pathways for bacterial mediated synthesis of noble metal nanoparticles. *Journal of Microbiological Methods* 159: 18–25. Doi: 10.1016/j.mimet.2019.02.010.

Aziz N., Fatma, T., Varma, A. and Prasad, R. (2014) Biogenic synthesis of silver nanoparticles using Tectona grandis leaf extract and evaluation of their antibacterial potential. *Journal of Nanoparticles* 6(1): 293–298.

Azmath, P., Baker, S., Rakshith, D. and Satish, S. (2016) Mycosynthesis of silver nanoparticles bearing antibacterial activity. *Saudi Pharmaceutical Journal. King Saud University* 24(2): 140–146. Doi: 10.1016/j.jsps.2015.01.008.

Banerjee, J. (2011) (L.) Seed extract and evaluation of their in vitro antioxidant activities. *Digest Journal of Nanomaterials and Biostructures* 6(3): 961–968.

Baskar, G., Chandhuru, J., Fahad, K. S. and Praveen, A. S. (2013) Mycological synthesis, characterization and antifungal activity of zinc oxide nanoparticles. *Asian Journal of Pharmacy and Technology* 3(4): 142–146.

Beveridge, T. J. and Murray, R. G. E. (1980). Sites of metal deposition in the cell wall of *Bacillus subtilis*. *Journal of Bacteriology* 141(2): 876–887. Doi: 10.1128/jb.141.2.876-887.1980.

Cannon, A. S. and Warner, J. C. (2011) The science of green chemistry and its role in chemicals policy and educational reform. *New Solutions* 21(3): 499–517. doi: 10.2190/NS.21.3.m.

Casadevall, A. and Pirofski, L. A. (1999) Host-pathogen interactions: redefining the basic concepts of virulence and pathogenicity. *Infection and Immunity* 67(8): 3703–3713. Doi: 10.1128/iai.67.8.3703-3713.1999.

Castro-Longoria, E., Vilchis-Nestor, A. R. and Avalos-Borja, M. (2011) Biosynthesis of silver, gold and bimetallic nanoparticles using the filamentous fungus *Neurospora crassa*. *Colloids and Surfaces B: Biointerfaces* 83(1): 42–48. Doi: 10.1016/j.colsurfb.2010.10.035.

Collins, C. Depledge, M., Fraser, R., Johnson, A., Hutchison, G., Matthiessen, P., Murphy, R., Owens, S. and Sumpter, J. (2020) Key actions for a sustainable chemicals policy. *Environment International* 137: 105463. Doi: 10.1016/j.envint.2020.105463.

Costa Silva, L. P., Oliveira, J.P., Keijok, W. J., da Silva, A. R., Aguiar, A. R., Guimarães, M. C. C., Ferraz, C. M., Araújo, J. V., Tobias, F. L., and Braga, F. R. (2017) Extracellular biosynthesis of silver nanoparticles using the cell-free filtrate of nematophagous fungus *Duddingtonia flagrans*. *International Journal of Nanomedicine* 12: 6373–6381. doi: 10.2147/IJN.S137703.

Das, R. K., Laxman, V. and Linson, P. (2017) Biological synthesis of metallic nanoparticles : plants, animals and microbial aspects. *Nanotechnology for Environmental Engineering* 2(1): 1–21. doi: 10.1007/s41204-017-0029-4.

David, L., Moldovan, B., Vulcu, A., Olenic, L., Perde-Schrepler, M., Fischer-Fodor, E., Florea, A., Crisan, M., Chiorean, I., Clichici, S. and Filip, G.A. (2014) Green synthesis, characterization and anti-inflammatory activity of silver nanoparticles using European black elderberry fruits extract. *Colloids and Surfaces B: Biointerfaces* 122: 767–777. Doi: 10.1016/j.colsurfb.2014.08.018.

Dias, M. A., Lacerda, I. C. A., Pimentel, P. F., de Castro, H. F. and Rosa, C. (2002) Removal of heavy metals by an *Aspergillus terreus* strain immobilized in a polyurethane matrix. *Letters in Applied Microbiology* 34(1): 46–50. Doi: 10.1046/j.1472-765X.2002.01040.x.

Dozie-Nwachukwu, S. O., Obayemi, J. D., Danuo, Y., Anuku, N., Odusanya, O.S., Malatesta, K. and Soboyejo, W. O. (2017) A comparative study of the adhesion of biosynthesized gold and conjugated gold/prodigiosin nanoparticles to triple negative breast cancer cells. *Journal of Materials Science: Materials in Medicine* 28(9). Doi: 10.1007/s10856-017-5943-2.

Dwivedi, A. D. and Gopal, K. (2010) Biosynthesis of silver and gold nanoparticles using Chenopodium album leaf extract. *Colloids and Surfaces A: Physicochemical and Engineering Aspects* 369(1–3): 27–33. Doi: 10.1016/j.colsurfa.2010.07.020.

Elegbede, J. A. and Lateef, A. (2019) Green nanotechnology in Nigeria: the research landscape, challenges and prospects. *Annals of Science and Technology* 4(2): 6–38. Doi: 10.2478/ast–2019-0008.

Ghosh, S., Patil, S., Ahire, M., Kitture, R., Kale, S., Pardesi, K., Cameotra, S. S., Ballare, J., Dhavale, D. D., Jabgunde, A. and Chopade, B. A. (2012) Synthesis of silver nanoparticles using *Dioscorea bulbifera* tuber extract and evaluation of its synergistic potential in combination with antimicrobial agents. *International Journal of Nanomedicine* 7: 483–496.

Gour, A. and Jain, N. K. (2019) Advances in green synthesis of nanoparticles. *Artificial Cells, Nanomedicine and Biotechnology* 47(1): 844–851. Doi: 10.1080/21691401.2019.1577878.

Gudikandula, K., Vadapally, P. and Singara Charya, M. A. (2017) Biogenic synthesis of silver nanoparticles from white rot fungi: their characterization and antibacterial studies. *OpenNano*. 2 (July): 64–78. Doi: 10.1016/j.onano.2017.07.002.

Hamelian, M., Zangeneh, M. M., Amisama, A. and Varmira, K. (2018) Green synthesis of silver nanoparticles using thymus kotschyanus extract and evaluation of their antioxidant, antibacterial and cytotoxic effects. *Applied Organometallic Chemistry* 32(9): 1–8. Doi: 10.1002/aoc.4458.

Hayyan, M., Hashim, M. A. and Alnashef, I. M. (2016) Superoxide ion : generation and chemical implications. *Chemical Reviews* 116: 3029–3085. Doi: 10.1021/acs.chemrev.5b00407.

Iravani, S. (2011) Green synthesis of metal nanoparticles using plants. *Green Chemistry* 13: 2638–2650. Doi: 10.1039/c1gc15386b.

Jacob, S.J.P., Finub, J. S. and Narayanan, A. (2012) Synthesis of silver nanoparticles using Piper longum leaf extracts and its cytotoxic activity against Hep-2 cell line. *Colloids and Surfaces B: Biointerfaces* 91(1): 212–214. Doi: 10.1016/j.colsurfb.2011.11.001.

Jamdagni, P., Khatri, P. and Rana, J. S. (2018) Green synthesis of zinc oxide nanoparticles using flower extract of Nyctanthes arbor-tristis and their antifungal activity. *Journal of King Saud University - Science. King Saud University* 30(2): 168–175. Doi: 10.1016/j.jksus.2016.10.002.

Jayaseelan, C., Rahuman, A.A., Kirthi, A.V., Marimuthu, S., Santhoshkumar, T., Bagavan, A., Gaurav, K., Karthik, L. and Bhaskara Rao, K.V. (2012) Novel microbial route to synthesize ZnO nanoparticles using *Aeromonas hydrophila* and their activity against pathogenic bacteria and fungi. *Spectrochimica Acta - Part A: Molecular and Biomolecular Spectroscopy* 90: 78–84. Doi: 10.1016/j.saa.2012.01.006.

Jumbam, N. D. and Jumbam, N. D. (2015) Transactions of the royal society of South Africa green chemistry in Africa – its inception and challenges green chemistry in Africa – its inception and challenges. *Transactions of the Royal Society of South Africa.* 70(2): 187–190. Doi: 10.1080/0035919X.2015.1019386.

Kojom Foko, L. P., Eya ane Meva, F., Eboumbou Moukoko, C.E., Ntoumba, A.A., Nagha Njila, M.I., Ebanda Kedi, P.B., Ayong, L. and Lehman, L.G. (2019) A systematic review on anti-malarial drug discovery and antiplasmodial potential of green synthesis mediated metal nanoparticles: overview, challenges and future perspectives. *Malaria Journal. BioMed Central* 18(1): 1–14. Doi: 10.1186/s12936-019-2974-9.

Lan, C. and Bao, Z. (2019) Light-dependent biosynthesis of silver nanoparticles mediated by cell extract of neochloris oleoabundans: an investigation on mechanism. *Proceedings of the 4th World Congress on Recent Advances in Nanotechnology. Environmental Science and Pollution Research* 26: 2873–2881. Doi: 10.11159/icnnfc19.2.

Larayetan, R., Ojemaye, M.O., Okoh, O.O. and Okoh, A.I. (2019) Silver nanoparticles mediated by *Callistemon citrinus* extracts and their antimalaria, antitrypanosoma and antibacterial efficacy. *Journal of Molecular Liquids* 273: 615–625. Doi: 10.1016/j.molliq.2018.10.020.

Li, X. Xu, H. Chen, Z.S and Chen, G. (2011) Biosynthesis of nanoparticles by microorganisms and their applications. *Journal of Nanomaterials* Doi: 10.1155/2011/270974.

Mata, Y. N., Blazquez, M. L., Ballester, A., Gonzalez, F. and Munoz, J.A. (2008) Characterization of the biosorption of cadmium, lead and copper with the brown alga *Fucus vesiculosus. Journal of Hazardous Materials* 158(2–3): 316–323. Doi: 10.1016/j.jhazmat.2008.01.084.

Mata, Y. N., Torres, E., Blazquez, M. L., Ballester A., Gonzalez F. and Munoz J. A. (2009) Gold(III) biosorption and bioreduction with the brown alga *Fucus vesiculosus. Journal of Hazardous Materials* 166(2–3): 612–618. Doi: 10.1016/j.jhazmat.2008.11.064.

Mohamed El-Rafie, H. and Abdel-Aziz Hamed, M. (2014) Antioxidant and anti-inflammatory activities of silver nanoparticles biosynthesized from aqueous leaves extracts of four terminalia species. *Advances in Natural Sciences: Nanoscience and Nanotechnology* 5(3). Doi: 10.1088/2043-6262/5/3/035008.

Mohd Yusof, H., Mohamad R., Zaidan U. H. and Rahman N. A. A. (2019) Microbial synthesis of zinc oxide nanoparticles and their potential application as an antimicrobial agent and a feed supplement in animal industry: a review. *Journal of Animal Science and Biotechnology* 10(1): 1–22. Doi: 10.1186/s40104-019-0368-z.

Molnár, Z. Bodai, V., Szakacc, G., Erdelyi, B., Fogarassy, Z., Safran, G., Varga, T., Konya, Z., Toth-Szeles, E., Szucs, R. and Lagzi, I. (2018) Green synthesis of gold nanoparticles by thermophilic filamentous fungi. *Scientific Reports* 8(1): 1–12. Doi: 10.1038/s41598-018-22112-3.

Mukherjee, P., Roy, M., Mandal, B. P., Dey, G. K., Mukherjee, P. K., Ghatak, J., Tyagi, A. K. and Kale, S. P. (2008) Green synthesis of highly stabilized nanocrystalline silver particles by a non-pathogenic and agriculturally important fungus *T. asperellum*. *Nanotechnology* 19(7). Doi: 10.1088/0957-4484/19/7/075103.

Nagajyothi, P. C., Pandurangan, M., Kim, A. H. and Sreekanth, T. V. M. (2017) Green synthesis of iron oxide nanoparticles and their catalytic and in vitro anticancer activities. *Journal of Cluster Science* 28(1): 245–257. Doi: 10.1007/s10876-016-1082-z.

Ojemaye, M. O., Okoh, O. O. and Okoh, A. I. (2017) Efficient adsorbents for heavy metal removal from wastewater. *Progress and prospects* 7(6): 439–456. Doi: 10.1166/mex.2017.1401.

Ojemaye, M. O., Okoh, S. O. and Okoh, A. I. (2021) Silver nanoparticles (AgNPs) facilitated by plant parts of *Crataegus ambigua* Becker AK extracts and their antibacterial, antioxidant and antimalarial activities. *Green Chemistry Letters and Reviews* 14(1): 49–59. Doi: 10.1080/17518253.2020.1861344.

Okaiyeto, K., Hoppe, H. and Okoh, A. I. (2020) Plant-based synthesis of silver nanoparticles using aqueous leaf extract of salvia officinalis: characterization and its antiplasmodial activity. *Journal of Cluster Science* 0123456789: 1–9. Doi: 10.1007/s10876-020-01766-y.

Okaiyeto, K., Ojemaye, M. O., Hoppe, H., Mabinya, L. V. and Okoh, A. I. (2019) Phytofabrication of silver/silver chloride nanoparticles using aqueous leaf extract of Oedera genistifolia: characterization and antibacterial potential. *Molecules* 24(23). Doi: 10.3390/molecules24234382.

Rotimi, L., Ojemaye, M. O., Okoh, O. O, Sadimenko, A. and Okoh, A. I. (2019) Synthesis, characterization, antimalarial, antitrypanocidal and antimicrobial properties of gold nanoparticle. *Green Chemistry Letters and Reviews* 12(1): 61–68. Doi: 10.1080/17518253.2019.1569730.

Sabri, M. A., Umer, A., Awan, G. H., Hassan, M. F. and Hasnain, A. (2016) Selection of suitable biological method for the synthesis of silver nanoparticles. *Nanomaterials and Nanotechnology* 6(1): 1–20. Doi: 10.5772/62644.

SACI. (2014) SACI office : KZN address advertising in the SACI newsletter SACI scarves and ties. *SACI*: 1–18.

Sahayaraja, K. and Rajesh, A. J. M. R. (2012) Silver nanoparticles biosynthesis using marine alga padina pavonica (linn.) and its microbicidal activity. *Digest Journal of Nanomaterials and Biostructures* 7(4): 1557–1567.

Saipriya, K. and Daisy, P. (2012) Biochemical analysis of *Cassia fistula* aqueous extract and phytochemically synthesized gold nanoparticles as hypoglycemic treatment for diabetes mellitus. *International Journal of Nanomedicine* 7: 1189–1202.

Salem, S. S. and Fouda, A. (2021) Green synthesis of metallic nanoparticles and their prospective biotechnological applications: an overview. *Biological Trace Element Research* 199(1): 344–370. Doi: 10.1007/s12011-020-02138-3.

Sau, T. K. and Murphy, C. J. (2004) Room temperature, high-yield synthesis of multiple shapes of gold nanoparticles in aqueous solution. *Journal of the American Chemical Society*, 126(28): 8648–8649. Doi: 10.1021/ja047846d.

Schwager, P., Decker, N. and Kaltenegger, I. (2016) Exploring green chemistry, sustainable chemistry and innovative business models such as chemical leasing in the context of international policy discussions. *Current Opinion in Green and Sustainable Chemistry* 1: 18–21. Doi: 10.1016/j.cogsc.2016.07.005.

Shivaji, S., Madhu, S. and Singh, S. (2011) Extracellular synthesis of antibacterial silver nanoparticles using psychrophilic bacteria. *Process Biochemistry* 46(9): 1800–1807. Doi: 10.1016/j.procbio.2011.06.008.

Singh, B. N. Rawat, A. K. S., Khan, W., Naqvi, A. H. and Singh, B. R. (2014) Biosynthesis of stable antioxidant ZnO nanoparticles by Pseudomonas aeruginosa Rhamnolipids. *PLoS One* 9(9). Doi: 10.1371/journal.pone.0106937.

Singh, J., Dutta, T., Kin, K. K., Rawat, M., Samddar, P. and Kumar, P. (2018) 'Green' synthesis of metals and their oxide nanoparticles: applications for environmental remediation. *Journal of Nanobiotechnology* 16(1): 1–24. Doi: 10.1186/s12951-018-0408-4.

Swarnalatha L., Christina, R., Shruti, R. and Payas, B. (2012) Evaluation of *invitro* antidiabetic activity of sphaeranthus amaranthoides silver nanoparticles. *International Journal of Nanomaterials and Biostructures* 2(3): 25–29.

Teske, S. S. and Detweiler, C. S. (2015) The biomechanisms of metal and metal-oxide nanoparticles' interactions with cells. *International Journal of Environmental Research and Public Health* 12(2): 1112–1134. Doi: 10.3390/ijerph120201112.

Thiruvengadam, M., Chung, I., Gomathi, T., Ansari, A. A., Khanna, V.G., Babu, V. and Rajakumar, G. (2019) Synthesis, characterization and pharmacological potential of green synthesized copper nanoparticles. *Bioprocess and Biosystems Engineering* 42(11): 1769–1777. Doi: 10.1007/s00449-019-02173-y.

Tippayawat, P., Phromviyo, N., Boueroy, P. and Chompoosor, A. (2016) Green synthesis of silver nanoparticles in aloe vera plant extract prepared by a hydrothermal method and their synergistic antibacterial activity. *PeerJ* 2016(10): 1–15. Doi: 10.7717/peerj.2589.

Tripathi, R. M., Bhadwal, A.S., Gupta, R.K., Singh, P., Shrivastav, A. and Shrivastav, B.R. (2014) ZnO nanoflowers: novel biogenic synthesis and enhanced photocatalytic activity. *Journal of Photochemistry and Photobiology B: Biology* 141: 288–295. Doi: 10.1016/j.jphotobiol.2014.10.001.

Ullah, H., Wilfred, C. D. and Shaharun, M. S. (2019) Green synthesis of copper nanoparticle using ionic liquid-based extraction from Polygonum minus and their applications. *Environmental Technology (United Kingdom)* 40(28): 3705–3712. Doi: 10.1080/09593330.2018.1485751.

UNIDO. (2018) *The Role of Technology and Innovation in Inclusive and Sustainable Industrial Development*. Doi: 10.18356/1b194elc-en.

United States Environmental Protection Agency. (2020) Basics of Green Chemistry. US EPA, Washington, DC.

Varadavenkatesan, T., Vinayagam, R. and Selvaraj, R. (2017) Structural characterization of silver nanoparticles phyto-mediated by a plant waste, seed hull of Vigna mungo and their biological applications. *Journal of Molecular Structure* 1147: 629–635. Doi: 10.1016/j.molstruc.2017.07.002.

Velusamy, P., Kumar, G. V., Jeyanthi, V., Das, J. and Pachaiappan, R. (2016) Bio-inspired green nanoparticles: synthesis, mechanism, and antibacterial application. *Toxicological Research* 32(2): 95–102. Doi: 10.5487/TR.2016.32.2.095.

Viscido, A., Capannolo, A., Latella, G., Caprilli, R. and Frieri, G. (2014) Nanotechnology in the treatment of inflammatory bowel diseases. *Journal of Crohn's and Colitis. European Crohn's and Colitis Organisation* 8(9): 903–918. Doi: 10.1016/j.crohns.2014.02.024.

Wang, G., Ahmeda, A. and Malek, Z. (2020) Chemical characterization and therapeutic properties of Achillea biebersteinii leaf aqueous extract synthesized copper nanoparticles against methamphetamine-induced cell death in PC12: a study in the nanotechnology and neurology fields. *Applied Organometallic Chemistry* 34(4): 1–14. Doi: 10.1002/aoc.5488.

Wang, L., Xu, J., Yan, Y., Liu, H., Karunakaran, T. and Li, F. (2019) Green synthesis of gold nanoparticles from Scutellaria barbata and its anticancer activity in pancreatic cancer cell (PANC-1). *Artificial Cells, Nanomedicine and Biotechnology* 47(1): 1617–1627. Doi: 10.1080/21691401.2019.1594862.

Wang, P., Lombi, E., Zhao, F. and Kopittke, P.M. (2016) Nanotechnology: a new opportunity in plant sciences. *Trends in Plant Science* 21(8): 699–712. Doi: 10.1016/j.tplants.2016.04.005.

Yadi, M., Mostafavi, E., Saleh, B., Davaran, S., Aliyeva, I., Khalilov, R., Nikzamir, M., Nikzamir, N., Akbarzadeh, A., Panahi, Y. and Milani, M. (2018) Current developments in green synthesis of metallic nanoparticles using plant extracts: a review. *Artificial Cells, Nanomedicine and Biotechnology* 46(suppl 3): S336–S343. Doi: 10.1080/21691401.2018.1492931.

Yin, Y., Yang X., Hu, L., Tan, Z., Zhao, L., Zhang, Z., Liu, J. and Jiang, G. (2016) Superoxide-mediated extracellular biosynthesis of silver nanoparticles by the fungus Fusarium oxysporum. *Environmental Science and Technology Letters* 3: 160–165. Doi: 10.1021/acs.estlett.6b00066.

Yurkov, A. M., Kemler, M. and Begerow, D. (2011) Species accumulation curves and incidence-based species richness estimators to appraise the diversity of cultivable yeasts from beech forest soils. *PLoS One* 6(8). Doi: 10.1371/journal.pone.0023671.

4 Enhancement of Bioprocesses for the Reclamation of Impacted Environment

Beckley Ikhajiagbe and Geoffrey O. Anoliefo
University of Benin

Francis Aibuedefe Igiebor
University of Benin
Wellspring University

Saheed Ibrahim Musa
Admiralty University of Nigeria

CONTENTS

4.1 INTRODUCTION

Pushpanathan et al. (2014) define pollution as the build-up and also the accumulation of pollutants within the water, land and air; this decreases the ability of contaminated sites to sustain life. Pollution may be a major problem all-round the globe, because it adversely affects immeasurable people, plants and animals per annum, causing

DOI: 10.1201/9781003180289-4

many health disorders and deaths. Universally, over one billion pounds of pollutants are discharged into the water and air. Pollution of soils with heavy metals and metalloids, for example, lead, cadmium and arsenic, speaks to a genuine risk for the biological system and human health. There is a critical requirement for remediation of this pollution so as to forestall its conceivable long-term impacts. It is therefore that certain moves are made to reduce the presence and impacts of contamination in the environment.

The utilization of living organisms, for example, fungi, microorganisms and plants, decreases the amount of contaminations, be it organic or inorganic toxin in the environment. The amount of pollutants can be reduced by organic procedures, for example, degradation and change of the contaminations into their non-dangerous forms. The particular methods by which bioremediation are accomplished are known as bioremediation advancements. Thus, bioremediation advancements refer to the different strategies applied in the treatment of pollutants in the environment. A few instances of bioremediation innovations that will be examined later on are: microbial remediation (the utilization of microorganisms), mycoremediation (the utilization of fungal organisms), phytoremediation (the utilization of plants), vermiremediation (utilization of earth worms) and synergistic bioremediation (the mix of at least two different advancements).

4.2 POLLUTION ABATEMENT STRATEGIES

Throughout the years, different measures have been applied in the interest to combat contamination in the environment and a portion of the measures utilized both previously and as of now will be talked about later on. There are various classes of contamination abatement methodologies: legitimate, social, economic and technological, yet our attention will be on the innovative techniques. There are different remediation advancements; however, they can be ordered depending on the area of the site: ex situ (the remediation happens off the site) and in situ (this spots on the site).

So as to acknowledge and comprehend the current technological remediation measures, some of the different measures from the past till now will be examined. Soil washing as a technique is used for removing heavy metals and organic pollutants from soils. Though it is not suitable for the cleanup of polluted soils, it is utilized in the mineral preparing industry. The fundamental standard of this procedure is a specific characterization of exceptionally polluted fines followed by the solid/liquid stage separation of the rest of the suspension. In spite of the fact that soil washing doesn't attack the pollutants, it isolates diverse soil parts with high contaminant content (Felix et al., 2020). Musa et al. (2019) had the option to remediate a soil polluted with 10% crude oil in the laboratory, where it was seen that the measure of heavy metals in the soil, also decreased.

A much more conventional remediation approach involves the haulage of the polluted soil to an observed landfill or air circulation of the uncovered material on account of unpredictable natural mixes. This is regarded as soil excavation or dredging. The pump-and-treat method is another conventional remediation approach that manages siphoning contaminated groundwater to the surface and decontaminating the extricated groundwater by permitting it to gradually go through a progression of vessels that contain materials intended to absorb the contaminants. The groundwater can likewise be treated via air-stripping, enacted carbon, bright or ozone treatment. Despite the fact that it might be a decent technique to lessen high measures of contamination (Zhang, 2005), this technique can be utilized to recuperate non-aqueous phase liquid.

Reddi and Inyang (2000) describe yet another procedure, whereby materials are changed over into glass or glass-like substances. This is called Vitrification. Other remediation methods include chemical oxidation (Nwachukwu, 2014) of contaminants with chemical agents like permanganate, persulfate and hydrogen peroxide. The utilization of concoction oxidants is profoundly reliant on the site; this is utilized for treating contaminants like amines, phenols, cyanides and certain pesticides in fluid waste streams. Some contaminants are encapsulated (Bradl and Xenidis, 2005), while others are simply burned. These methods have their pitfall; most of the time the contaminants may not be entirely removed from the environment. In some other cases, by-products of the use of most chemical oxidation of organic contaminants may themselves pose some level of environmental threats. Burning of contaminants creates carbon smoke, which itself is a contaminant. The application biological processes for the removal of contaminants are perhaps a much more ecofriendly measure. This is called bioremediation.

Bioremediation incorporates each one of those procedures and activities that utilizes living life forms to restore nature modified by contaminants to its unique condition (Garbisu and Alkorta, 2003). It might be utilized so as to assault explicit contaminants, for example, chlorinated pesticides, oil slicks and heavy metals. It principally manages the utilization of microorganisms, organisms, green plants or their compounds to degrade and change ecological contaminants into innocuous or less lethal structures (Jobby and Desai, 2017). The utilization of Indian mustard plant by Edenspace Systems Corporation to treat soil contaminated by radionuclide strontium ($Sr^{89/90}$) at Fort Greely in Alaska, USA, and Cs^{137} from the tainted lake waters is an example (Singh et al., 2006).

4.3 BIOREMEDIATION TECHNOLOGIES

The usage of organic substances in the recovery of contaminated condition is alluded to as bioremediation. The development of bioremediation advances for ecological cleanup stems from the environment's innate ability to self-clean up, a process known as natural attenuation. In the offer to self-remediate, three different degrees of remediation are incorporated in guaranteeing alongside ideal tidy up. Physical remediation is generally one of initial steps to characteristic weakening. It is the upgrade for temperature and other physical ecological properties that improve decrease of focus or degradation of complex contaminants. Chemical remediation during normal weakening of dirtied condition may incorporate the synthetic procedures like redox responses occasioned by common segments of the situations in soil, air and water, though bioremediation, as prior suggested, includes all procedures required by organic elements and sentinels in the environment.

Bioremediation can be arranged dependent on where remediation happens or dependent on the life forms utilized in the remediation. In view of the area where the remediation happens, bioremediation could be in situ when the bioremediation procedure is completed on-site or ex situ when such action is done off-site. During ex situ remediation, the contaminated substrate is expelled from site of tainting to site of remediation and afterward returned after tidy up. Be that as it may, in view of the kind of living being associated with the remediation, bioremediation can be ordered into microbial remediation when the organisms used are microorganisms, especially individuals from the Kingdom Monera and Protista. Mycoremediation or fungi bioremediation subsists when the organisms utilized are individuals from the Kingdom Fungi. The utilization of members of the Kingdom Plantae or the utilization of plants in the recovery of contaminated conditions is referred to as phytoremediation. Different classifications happen under the terminology Miscellaneous Remediation; a model is vermiremediation which uses worms during remediation.

4.4 MICROBIAL BIOREMEDIATION

Microorganisms are minute living beings that exist as unicellular, multicellular or cell groups. Microorganisms are differing and are found in the three areas of life: Archaea, Bacteria and Eukarya (Lumenlearning.com, 2020). Despite the fact that archaea can likewise be utilized in microbial bioremediation of extraordinary conditions, for example, halophilic or acidophilic situations (Krzmarzick et al., 2018), our attention will be on the prokaryotes, explicitly the microbes.

Bacteria exist in all aspects of the environment because of their shifted nourishing and metabolic limit which additionally makes them key players in the guideline of biogeochemical cycles in the different natural conditions running from solidified situations, aqueous vents, acidic lakes, seas and deserts to even the guts of creatures (Das and Dash, 2014), which empowers them in the breakdown of waste. As per Pushpanathan et al. (2014), due to the tremendous hereditary and metabolic assorted variety, microorganisms additionally have an incredible potential for their application toward biotechnology purposes. It is for this reason close by their flexibility and versatility to different testing conditions (Das and Dash, 2014) that microorganisms are every now and again applied in the remediation of pollutants. Consequently, microbial bioremediation is the procedure including the biodegradation of pollutants by the movement of microbial compounds, which prompts the bioaccumulation or bioconversion of lethal contaminations to non-pollutants or less dangerous substances (Pushpanathan et al., 2014; Osarumwense and Igiebor, 2018). It gives successful intends to tidying up the contaminations present, for instance, degradation of pollutants to supplements which can be utilized by the microorganisms and plants also (Osarumwense and Igiebor, 2018). Aside from the degradation of pollutants, microscopic organisms additionally render contaminations bioavailable for different organisms so as to expand the remediation procedure. A few microscopic organisms utilized in microbial bioremediation are: *Arthrobacter* spp., *Pseudomonas veronii*, and *Moraxella* and *Bacilli*. There are different methods created to improve the interactions between the microorganisms and the contamination (Das and Dash, 2014).

Bioaugmentation includes the presentation of a gathering of characteristic microbial strains or hereditarily designed variation to take care of polluted water or soil. It includes gathering of microorganisms like bacteria, protozoa, nematodes, rotifers and fungi, fit for debasing natural mixes. Biostimulation is used to treat subsurface area by expansion of water-based arrangements conveying nutrients, electron acceptor or different changes. This strategy is utilized to treat soil and groundwater tainting by fuels, non-halogenated unpredictable natural mixes and semi-unstable natural compounds, pesticides and herbicides.

Another improved strategy includes soil vapor extraction. This is powerful at tidying up source zones containing unpredictable mixes in homogeneous porous soils. One of the most recognized in situ treatment methods that is planned essentially to take care of soil pollution by powers, unpredictable natural mixes and semi-unstable natural compounds, pesticides and herbicides is bioventing. It includes oxygen conveyance as air to contaminated unsaturated soils through an arrangement of extraction and infusion wells. Like bioventing is air sparging. Air sparging is utilized on the contaminated groundwater by infusing air beneath the water table. It is planned essentially to treat groundwater pollution by fuels, non-halogenated unpredictable natural mixes and semi-unstable natural compounds, pesticides and herbicides.

Land cultivating is additionally a technique for upgrading bioremediation of contaminated conditions: Here, tainted soil is dug out and spread over a readied platform/bed and occasionally worked to accomplish the breakdown of pollutants. Furthermore, microorganisms encourage oxygen consuming breakdown of pollutants. Every day, contaminants are discharged into the environment, deliberately or accidentally, which affect the natural functioning of ecosystem. Henceforth, there is the need to remediate these pollutants, yet with pollutants being either organic or inorganic; one must comprehend the instrument by which the two kinds of contaminations are tidied up individually so as to successfully do bioremediation. Hence, we will be taking a look at the instrument of organic pollutant tidy up. Firstly, when organic pollutants such as polycyclic and other aromatic hydrocarbons, halogenated compounds and petroleum products, pesticides and synthetic dyes interact with the microbial enzymes, the action and the portability of the pollutants become confined because of the catabolic properties of the enzyme, and the lowered activation energy of the reaction. With respect to insoluble contaminants unlike to soluble pollutants, they must be changed to less complex units before they can be taken into the bacterial cell. This is finished by extracellular proteins which incorporate an expansive range of oxidoreductases and hydrolases having degradative capacity which permits them to separate natural toxin into mostly oxidized which can be promptly taken into the cell to experience total mineralization. The results of this procedure have expanded polarity and water solubility and expanded bioavailability (Meulenberg et al., 1997).

The remediation of inorganic contaminants follows a lot more different approach from organic contaminant cleanup. Here, heavy metals, metalloids and radionuclides which can't be corrupted are changed over into some structure with the assistance of metal-tolerant microorganisms. In circumstances where microorganisms are utilized in mix with another life form, for example, plants for instance, the microbes follow up on the inorganic toxin making it bioavailable which builds the plants productivity in collecting and settling the contamination (Ahmad, 2017).

Under typical conditions when the microbes work alone, the microorganisms either advance bioaccumulation of the inorganic contaminations or advance the adjustment of the metal in the soil by discharging natural mixes, for example, oxalic corrosive and citrus extract which bind to the metals. Microscopic organisms can evacuate metals, for example, Arsenic through methylation, volatilization and redox responses. Microscopic organisms additionally can change, solubilize and gather radionuclides. Jobby and Desai (2017) expressed that microorganisms through retention take in inorganic contaminations into their cell wall components. Microorganisms can biodegrade contaminations by biosorption which is a passive procedure that happens in either living and dead biomass or bioaccumulation, a functioning procedure that happens in living biomass only (Coelho et al., 2015).

Various confinements exist, depending upon the lethality of the pollutant, a high sum may have adverse effects on the microbial metabolism. Likewise, degradation may not happen if the measure of the pollutants is too low. There are contaminations that can't be remediated by known bacteria. There are situations where the microorganism changes over the less destructive pollutant into an increasingly harmful product. Microorganisms work at certain scope of temperature, supplements accessibility and dampness content in addition to other things, so when these conditions are not met, their degradation is impeded. In dirtied springs, protists brush on contamination degrading microbial networks which can affect the general biodegradation rates (Vogt and Richnow, 2013). The pollutants may not be bioavailable because of biosorption to solids, nearness in nonaqueous stage lipid or capture inside physical framework of the environment.

4.5 MYCOREMEDIATION

Mycoremediation is characterized as the way toward detoxifying environmental contaminants. This is one of the systems of bioremediation. Mycoremediation refers to any technique for detoxifying pollutants utilizing fungi. In mycoremediation, fungi disengaged from nature might be tried to decide if they are fit for breaking down the pollutant, and subsequently investigating autochthonous pollutant-debasing microorganisms which could uncover if bioaugmentation will be required. The capacity of fungi to separate fragrant and phenol hydrocarbons in wood has been known for a considerable length of time. During the 1960s, Dr. Horst Lyr depicted ligninolytic reduction by white decay fungi (Christin and Glenn, 2016). White decay organisms are wood decomposers that specially debase lignin, abandoning thick, white cellulose and hemicellulose. Bumpus (Christin and Glenn, 2016) developed these underlying discoveries, theorizing that white spoil fungi possibly will debase constant xenobiotics, which have carbon skeletons like lignin. Bumpus assessed the synthetic reduction of trinitrotoluene (TNT), polychlorinated biphenyls, lindane and dioxins, by *Phanerochaete chrysosporium*.

As examinations of mycoremediation have continued, the broadness of contaminations growths known to degrade has likewise expanded. Notwithstanding oil and polycyclic aromatic hydrocarbons (Igiebor et al., 2017), white rot fungi were additionally observed to break down pesticides, polychlorinated biphenyls, explosives, creosote and engineered colors (Pointing, 2001). Fungi that could live on the contaminant alone, with no co-digestion, were found surviving on unpredictable organic hydrocarbons and polyurethane (Russell et al., 2011). As reports of extra fungal species and contaminations they contaminated grew, another field rose, and the expression "mycoremediation" was instituted in 2005 by Paul Stamets.

Many oil-reducing bacteria have been isolated from the environment, and a number of them are more successful than microorganisms in totally removing crude oil (Christin and Glenn, 2016). Recent studies have shown that microfungi, for example, *Cladosporium* sp. (Giraud et al., 2001), *Penicillium* sp., *Ascomycetes* (Musa, 2019), and other *Deuteromycetes* isolated from the environment, can breakdown hydrocarbons (Prenafeta-Boldu et al., 2001; Igiebor et al., 2017). In any case, in about these investigations of fungal reduction or breakdown of petroleum products, the species analyzed were microfungi. However, most white rot fungi are *basidiomycetes*, which comprise of a mycelia that can pull water and supplements all through the substrate. Furthermore, the mycelia structure of basidiomycete permits the creature to connect holes between patches of contamination, rather than requiring persistent contact with the contaminant as a development substrate (Harms et al., 2011). A preferred position for contaminants that are widely dispersed is the ability to remedy the contamination.

Pleurotus ostreatus has numerous eye-catching features as species for mycoremediation. It is an adaptable white rot basidiomycete that has been demonstrated effective in degrading petroleum in numerous investigations (Leonardi et al., 2006; Meysami and Baheri, 2003; Okparanma et al., 2011; Zitte et al., 2012; Young et al., 2015). *P. ostreatus* generates large scope of proteins dynamic alongside a wide scope of mixes. *P. ostreatus* can ship water, supplements, microscopic organisms and the pollutants themselves along the mycelium, sharing assets for pollutant degradation all through the soil matrix. At last, *P. ostreatus* can traverse air-filled holes in soil and infiltrate micropores, and co-process unharmful natural particles, permitting the parasite to stay dynamic in conditions antagonistic to bacterial development (Wick et al., 2010).

Be that as it may, the use of *P. ostreatus* in mycoremediation has various weaknesses. In the first place, the growth requires a vigorous domain. Second, when built up, the mycelia are touchy to unsettling influence. Standard techniques for bioremediation that circulate air through by working are intended for microbes, and possibly harm the mycelium by separating it. At the end, *P. ostreatus* needs a lignocellulose change, since it is a wood rotting fungus, and may possibly contend well with local soil fungi and microorganisms with no rule substrate (Harms et al., 2011).

In analyzing mycoremediation potentials, a key inquiry is the manner in which fungi accomplish the degradation of contaminations. The regular ligninolytic chemicals in white rot fungi are peroxidases and laccases (Elisashavili et al., 2008). They are both extracellular proteins that breakdown hydrocarbon reduction utilizing free radicals, atoms with distinct unpaired valence electrons. Laccases have been known to oxidize oxygen to discharge a free radical, while peroxidases oxidize hydrogen peroxide, Mn^2 or Mn^3 to discharge free radicals (Harms et al., 2011). The free radicals unpredictably divide the bonds, and breaking straight chain hydrocarbons or aperture aromatic rings, in this manner making hydrocarbons increasingly bioavailable for take-up and biologically degraded by organisms, microbes and different degraders (Christin and Glenn, 2016). In fungi, a few metabolites are likewise taken into the hyphae and further degraded with cytochrome P450 (Harms et al., 2011). With their mixed drink of vague chemicals, white rot fungi are a portion of the main living organism ready to totally mineralize hydrocarbons (Musa, 2019).

The exhibition of peroxidases and laccases could be influenced by ecological environments. Lignin reduction seems, by all accounts, to be invigorated more as a reaction to supplement exhaustion than to the nearness of contaminant (Madhavi and Lee, 2009). Ligninolytic chemical creation is most noteworthy during the vegetative development stage of fungal and diminishes pointedly during fruiting. Laccase action can be expanded by optimum temperatures (Madhavi and Lee, 2009). Laccase creation could be restrained by low accessibility of copper, given that copper is a significant component of laccase (Tychanowitcz et al., 2006). However, not all laccases contain four copper atoms; one disconnected from *P. ostreatus* controlled one zinc atom, one copper atom, and two iron atoms rather (Christin and Glenn, 2016).

Inadequate mineralization can give harmful metabolites (Pinedo-Rivilla et al., 2009; Christin and Glenn, 2016). Quinones are regular metabolites of polycyclic aromatic hydrocarbons (PAHs), while some can be more harmful than the underlying compound. Ligninolytic chemicals can additionally mineralize quinones, or make them increasingly biologically available to reduce by microbes (Pozdnyakova, 2012). The intracellular cytochrome P_{450} catalysts are likewise ready to degrade into metabolites into totally mineralized water, carbon dioxide and proteins (Harms et al., 2011).

4.6 SYNERGY WITH BACTERIA

Albeit white spoil organisms can totally mineralize hydrocarbons, all the more normally they perform just the underlying assault, tearing release hydrocarbon fastens and rings to make the mixes progressively accessible to microbial decay. During a 60-day investigation of petroleum bioremediation, Chen et al. (2009) observed that *Bacillus subtilis* and *Glomus caledonium*, utilized collectively, evacuated 92.6% of absolute petroleum hydrocarbon while autochthonous microorganisms expelled just 21.9%, and the joined impact was more noteworthy than the aggregate of entity expulsion by each in pure culture. In certain species, the mycelia structure rhizomorphs can lead water all through the soil, conveying microscopic organisms to in any case distant regions (Kohlmeier et al., 2005). Li et al. (2008) watched a synergistic impact of *Mycobacterium* and filamentous growths, with diesel loss expanding from 20.9% and 34% independently to 99% together over a period of 5 days treatment.

4.7 PHYTOREMEDIATION

The biological system is more and more unprotected to changes coming about because of both natural and anthropogenic sources. These changes could be extraordinary and as such influence nature essentially. Natural contamination is dominating in Nigeria, because of the lifestyle of the individuals and being a significant oil-creating nation. Crude oil and its subsidiaries represent more than 90% of Nigeria's national income. Contamination activated by the refined items, for example, spent motor oil and diesel, has not been given the proper thought it justifies; the impacts of this spent released motor oil are ineffectively taken care of, as only restricted information exists on how the spent motor oil is to be used, and they are not even promptly accessible for use in nearby networks around the country.

Be that as it may, farmlands are in danger because of oil slicks, with sweeping effect on rural harvests. This demonstration as a rule causes unexpected negative and regularly extreme responses with interest for endurance by living things inside the biological system (plants, water bodies, animals and human). Musa (2019) revealed that contamination because of used motor oil is more predominant and more overpowering than crude oil contamination. This happens in light of the fact that it is a significant reason for contamination inside urban and rural communities, along these lines, posing a great danger to plants, animals, human health, diminished soil richness and microbial/fungi movement. Nigeria represents in excess of 87 million liters of WEO every year (Okparanma et al., 2011).

Oil pollution is a significant environmental dread in Nigeria, and this has prompted numerous thorough works in contemplating the practicability of utilizing different techniques for remediation. Remediation advancements are numerous and changed so over the span of this investigation, we will be seeing one significant fragment and its phytoremediation.

Phytoremediaton is the procedure whereby plants are utilized to tidy up the environment. The word phytoremediation originates from the Greek word *phyto*, signifying "*plant*," and the Latin word *remediare*, signifying "to remedy." This word is usually used to clarify any procedure whereby plants are acquainted with expel contaminants from a given situation. Phytoremediation has a wild range in application, the plants can be brought into a situation and permitted to assimilate contaminants into its leaves, stem and roots, break down the contaminating material, volatilize the contaminate and keep the pollutant inside itself. These plants would then be able to be gathered and treated as hazardous waste. A few studies have been done to demonstrate how these plants effectively transformed the contaminant into a harmless substance and afterward once gathered can be utilized for other agricultural procedures like mulch, animal feed, paper and so forth. In certain examples (developed trees are utilized), the plants are left in the local condition and allowed to develop and develop on the grounds that as they keep on outstanding, they are clearing the air and furthermore filling in as a carbon sink for the environment.

Ikhajiagbe et al. (2017) said that the capacity of *Celosia trigyna*, *Talinum triangulare*, *Corchorus olitorus*, *Vernonia amygdalina* and *Telferia occidentalis* just like grasses such as *Eleucine indica*, *Panicum maximum*, *Cynodon dactylon*, *Euphorbia hirta* and *Chromolaena odorata* to take up heavy metals from soil was available in the entirety of the above named plants (Wong and Chu, 1985; Wong and Lau, 1985; Ikhajiagbe et al., 2017).

Genetic engineering and breeding plants are powerful methodologies for improving natural limits of plant, or for bringing these tendencies into interchange methods for plant which may be increasingly reasonable for the ecological conditions.

One of the mechanisms of phytoremediation involves the accumulation of contaminants, particularly inorganic contaminants, in harvestable plant parts. This is called phytoaccumulation or phytoextraction. In order to be appropriate for the purpose of phytoextraction, plant species should meet the following criteria; it must be metallotolerant toward elements present in toxic levels. It also has to be capable of a high biomass production. Finally, being effective in the build-up of pollutant in easy-to harvest parts is a noteworthy characteristic of a would-be phytoaccumulator plant.

The general idea of the rebuilding of contaminated areas utilizing phytoextraction includes the development of legitimate plant species in situ, reaping the pollutant containing biomass and getting it decline its weight and volume (by treating the soil, drying, compacting, warm decay). Significant levels of metal contaminants are enhanced by the resultant pollutant mass and it is along these lines slanted of as a destructive waste or, if financially worthwhile, can be used for the re-extraction of trace elements. Rhizofiltration is the expulsion of dangerous metals from ground water.

Many plants lack the capacity to degrade some types of contaminants including pesticides, and in most cases heavy metals which cannot be accumulated by the plants in other not to become poisoned by the accumulated metal. The plants rely on a number of chemical agents called chelatins or organic ligands to bind with active sites of these contaminants, thus rendering them "incapacitated" around the plants. This is called phytostabilization. Phytostabilization can be improved by utilizing soil changes that immobilize metal(loid)s joined with plant species that are tolerant of significant levels of pollutants and low-richness soils or tailings. Despite the fact that this innovation is effective in the regulation of metal(loid)s, the site requires customary observing to guarantee that the balancing out conditions are kept up. Maybe soil amendments used to upgrade immobilization ought to be occasionally reapplied to keep up their adequacy. We fundamentally inspect the relevance of this innovation to oversee metal(loid)s contaminated soils and recognize ripe regions for future research.

Plant-microbial synergy is key to enhancing bioremediation successes in any contaminated environment. Through phytostimulation, a process where root discharged compounds that upgrade microbial action in the rhizosphere (root region), plants are able to enhance microbial association with a view to speeding up or enhancing the environmental reclamation processes. The procedure is crucial for the formative procedures for rhizoremediation which works with phytoremediation and bioaugmentation. This kind of rhizosphere phytoremediation is practical in light of the fact that, natural contaminations can be expelled effectively from the soil. Phytostimulation is key for rhizoremediation for different reasons. To start with, the rhizosphere gives a specific specialty to microorganisms in the soil to live. This specialty is continually developing as roots develop and invade new soil zones. Root exudates actuate different advantages for microorganism development and action. The compounds discharged by the root have numerous jobs in phytostimulation since they regularly impersonate man-made contaminations. The rhizosphere fills in as a remarkable soil living space, loaded with natural aggravates that may fill in as vitality hotspots for microorganisms. Then again, these natural mixes may fill in as inducers for microbial genes engaged with degradation, or as co-metabolites that are essential for pollutant degradation to happen. A great case of a plant auxiliary metabolite that can initiate a bacterial protein significant for PAH degradation is salicylate, which incites naphthalene oxygenase in *Pseudomonas*.

Extra part of phytostimulation is the creation of new microorganisms in the soil, because of the distinctive rhizosphere environment that is delivered by living plants; microorganisms are now and again vaccinated on plant seed to introduce explicit organisms to the developing rhizosphere. Organisms inoculated might be hereditarily designed or naturally happening strains that carryout specific purpose of interest for remediation.

An important mechanism of phytoremediation is phytotransformation or phytodegradation. This involves the breakdown of natural contaminants sequestered by plants through metabolic process inside the plant; or the excess compounds, for example, chemicals, made by the plant. The natural contaminants are separated into littler aggravates that are incorporated with plant tissue, which thus cultivates plant development. Remediation of a site by phytotransformation is reliant on direct take-up of contaminants from the media and aggregation in the vegetation. Direct take-up of synthetic compounds into plant tissue by means of the root framework is subject to take-up proficiency, transpiration rate and grouping of the substance in soil water.

Plants have the ability to withstand moderately high convergences of natural synthetic chemicals, for example, oil, without harmful impact, and they can take up and convert these contaminants rapidly to less dangerous metabolites at times by rhizodegradation. This is accomplished by the arrival of root exudates, proteins that invigorate mineralization of the oil contaminant. This procedure develops natural carbon in the quick soil condition. Leguminous plants, for example, *P. vulgaris* equipped for fixing nitrogen, may have had advantage in the remediation of oil-polluted soil on account of their capacity to give their own nitrogen manure by means of nitrogen obsession to address irregularity in C:N proportion.

In many cases, some of the degraded organic contaminants may produce volatile products. These products may eventually escape from the plants through openings such as stomata and lenticels. This process is called phytovolatilization. Phytovolatilization is frequently observed as beneficial, in light of the fact that the procedure results in by and large weakening of the contaminant and photochemical rot in the climate. However, phytovolatilization might be viewed as a hazard in enormous areas, where introduction potential exists and air quality is contaminated by the released compound.

4.8 SYNERGISTIC BIOREMEDIATION

Interactions among plants and microorganisms are a vital piece of our environment bound biological system. There are a few kinds of plant–microbe collaborations: rivalry, mutualism, commensalism and parasitism. The more typical interactions are mutualism or commensalism, where it is possible that one or the two animal types profit by the relationship individually. There are a few phenomenal audits announcing ebb and flow look into on ways of life and atomic connections of plant-related microscopic organisms (Sorensen and Sessitsch, 2007), rhizospheric communications (Singh et al., 2004), endophyte applications (Ryan et al., 2008), plant reactions to bacterial quorum-sensing (QS) signals (Musa, 2019) and rhizospheric microorganisms reactions to transgenic plants (Fillion, 2008).

Assessment of these interactions encourages us to comprehend characteristic marvels that influence our day by day lives and could prompt applications bringing about reasonable assets, less effect on the environment, cleanup of contamination and impact on barometrical gases on a worldwide scale. Focal points of utilizing these connections for biotechnological applications are many-overlap. The utilization of normally existing plant-microorganism advantageous interaction for plant development and biocontrol lessens engineered manure and pesticide medications prompting cost-viability and less effect by supplements and pesticides (Elmer and Reglinski, 2006) on encompassing fauna and vegetation. The creation of valuable mixes with pharmaceutical and technological pertinence utilizing plant–microbes advantageous interaction is vitality productive (Wu et al., 2007; Del Giudice et al., 2008) and lessens the need to include costly forerunners and impetuses. Remediation through regular strategy, for example, uncovers and treats, is costly and work concentrated. On the other hand, plant-microbial remediation procedures can be less nosy and considerably more prudent. This audit centers on advances in the fields of plant-growth advancement, for the remediation of polluted locales.

Plant-microorganism interactions have been used to enhance plant development for the creation of nourishment, fiber, biofuels and key metabolites. The mutualistic association can be advantageous in straightforwardly giving supplements to the plant (biofertilizer) or expanding the accessibility of compounds, for example, iron or phosphate. Free-living plant growth advancing microscopic organisms additionally produce exacerbates that legitimately influence plant digestion or balance phytohormone creation or reduction. The phytohormones, cytokinins, auxins, gibberellic corrosive (GA3), abscisic corrosive and ethylene, are flagging particles fundamental for development which intercede a scope of formative procedures in plants.

4.9 BIOCHEMISTRY OF DEGRADATION

The biochemical procedures that manage the degradation of pollutants include the change of the pollutant to littler weight sections that can be further debased as often as possible to carbon dioxide (CO_2) and water. Degradation can be completed by bacteria, fungi and plants, whereby it is referred to as biodegradation. In nature, bacteria and fungi are the principle degraders which are vital for the handling of polymers. With respect to pollutants being either natural or inorganic, inorganic pollutants are very impervious to biodegradation; however, the change of insoluble polymers to solvent polymers is made conceivable using these microorganisms. The solvent polymers once separated by the microbial proteins are used for energy creation.

The Tricaboxylic Acid (TCA) cycle fills in as one of the principle metabolic pathways for vitality age from the majority of the natural compounds. In the TCA cycle, acetyl-CoA acts a key middle of the road and is misused in cell exercises like CO_2 arrangement, biosynthesis and so on. The significant commitment of TCA cycle is to produce ATP and give vitality to the cell. Two particles of ATP are incorporated by oxidation of two atoms of acetic acid derivation, while 34 atoms of ATP is blend by the electron transport chain (ETC) through phosphorylation (Pathak and Navneet, 2017). Microbial degradation of polymers prompts the change of physicochemical properties of materials. The bioconversion procedure additionally changes the effectiveness of the host response, cellular growth and material capacity.

4.10 GENE INTERACTION AND EXPRESSION DURING REMEDIATION

Genes are units of heredity that are sectioned in DNA particles. They control the expression of characters in creatures. Genes responsible for coding of proteins which helps manage all activities of living organisms'slife, helping them work and develop. Genes essentially lead to the creation and articulation of everything in the organism. Mobile genetic elements, for example, plasmids and transposons, have been found to encode compounds liable for the reduction of a few pesticides. Glycosidic surfactant is encoded by *rhl* A, B, R and I genes. These microbially created surface-dynamic compounds have the capacity to decrease the surface and interfacial strain between two immiscible liquid stages.

It has been investigated by environmental microbiologist that the emission of the glycosidic surfactant, Rhaminolipids, is for the most part a piece of a normally evolved component for improved substrate take-up. The predicted possibilities of biosurfactants in bioremediation of oil pollution cannot be overlooked (Oberbremer et al. 1990; Finnerty 1994). Biosurfactants are of various chemical structures. They incorporate the lipopeptides (He et al., 2001), glycolipids (Robert et al., 1989; Morita et al., 2007), flavolipids, polymeric and particulate sorts and they are delivered by both fungi and bacteria (Vance-Harrop et al. 2003).

Diverse microbial catalysts with the capability to hydrolyze pesticides have been recognized (Yan et al, 2007), for example, organophosphorus hydrolase (OPH; encoded by the opd quality) (Ortiz-Hernândez et al., 2013). This gene has been found in bacterial strains that can utilize organophosphate pesticides as carbon source. The plasmids show significant hereditary decent variety, yet the district containing the *opd* gene is profoundly monitored (Ortiz-Hernândez et al., 2013). Methyl-parathion hydrolase (MPH; encoded by the mpd gene), are *Pseudaminobacter*, *Achrobacter*, *Brucella* and *Ochrobactrum* genes (Zhongli et al., 2001), the gene for the organophosphorus hydrolase has 996 nucleotides (Zhang et al., 2005).

A comprehension of the hereditary premise of the mechanism of how microorganisms biodegrade contaminants and how they interface with the environment is significant for effective usage of the innovation for in situ remediation (Hussain et al., 2018).

4.11 BIOTECHNOLOGICAL IMPROVEMENTS OF REMEDIATION PROCESSES

Industrialization and other anthropogenic exercises have brought about genuine natural contamination in the 20th century. All inclusive, environmental contamination of both land and water ways have expanded to a level that imperils human health and influence nourishment productivity and security. A few strategies for remediating and recovering this polluted environment have been proposed by different scientists (Musa et al., 2018). The synthetic and physical strategies demonstrated not to be ecofriendly and cost sparing and may bring about different types of pollution. For this reason, flow looks into has dependent on biotechnological approaches in remediation of harmful natural synthetic chemicals. Biotechnology manages different advances that attempt to utilization of living frameworks or organisms to take care of human and environmental issues. This is otherwise called bioremediation. It is a novel technique that uses microorganisms, plants and animals in expending or separation of organic pollutants so as to tidy up sites (Musa et al., 2018). This component may include degradation, detoxification, mineralization, change, bioaccumulation and extraction of contaminations from environment. As of late, there have been advances in bioremediation methods. Anyway because of nature and also because of kind of pollutant, there is no single bioremediation that fills in as the best to re-establish polluted environments. Novel (indigenous) organisms that are thriving in contaminated situations hold the way to solving the greater part of the difficulties related with bioremediation and biodegradation of contaminating substances (Verma and Jaiswal, 2016) given that ecological conditions are appropriate for their development and digestion.

Bioremediation system for the most part takes in light of the site and the natural conditions. In view of this, biotechnologist has arranged two systems which are ex situ and in situ (Frutos et al., 2012; Smith et al., 2015). Considering this, there are different factors, for example, oxygen and supplement fixations, pH, temperature, cation exchange limit, and other abiotic factors, that decide the success of bioremediation processes. Ex situ bioremediation includes remediation outside the contaminated condition. This depends on the depth of contamination, sort of pollutant, environmental area and geography of the soil (Philp and Atlas, 2005). These procedures incorporate biopiling, windrow, bioreactor use and land cultivating (Gomez and Sartaj, 2013; Khudur et al., 2015). In situ bioremediation methods then again include treating the contaminated environment nearby. This doesn't require development of the pollutants from its common habitat. For this reason, there is almost no unsettling influence of the environment and it's considered as more affordable contrasted with the ex situ. This technique can include bioventing, biosparing, natural attenuation, microbial remediation, vermiremediation and phytoremediation (Folch et al., 2013; Kim et al., 2014; Frascari et al., 2015; Roy et al., 2015). These techniques have demonstrated successful remediation. Be that as it may, with ever increment in ecological contamination because of anthropogenic exercises and industrialization, these methods have to be improved for progressively compelling remediation.

4.12 BIOTECHNOLOGICAL IMPROVEMENT

Diverse investigation has attempted to improve the distinctive remediation techniques in request to adequately recover the expanding environmental contamination and to enhance a portion of the confinement experienced in remediation. The technological presentation of novel hereditary tools and a superior comprehension of microorganism, animal and plant gene structures and capacities have quickened enhancements in biotechnological methodologies of environmental remediation. The present section will concentrate on how the current biotechnological techniques have been as of now improved for productive and viable environmental protection, detoxification and the expulsion of metalloids and heavy metals. Next age sequencing has been used to study the entire transcriptomes and genomes of a few heavy metal-tolerant organisms (Hu et al., 2005; He et al., 2011; Peña-Montenegro and Dussán, 2013). Mass spectrometry-based proteomics is widely used to concentrate heavy metal and different types of stresses in individual organisms including plants (Hossain and Komatsu, 2012; Cvjetko et al., 2014), bacteria (Zakeri et al., 2012) and marine organisms (Muralidharan et al., 2012). Diverse biotechnological upgrades methodologies are currently being utilized to improve microbial remediation, phytoremediation and bioaccumulation.

Microbial remediation essentially manages the utilized microorganisms to eliminate environmental pollutants from a specific environment (Naz et al., 2015). Utilizing microorganisms for remediation ends up being a viable methodology because of their differed capacity of associating with pollutants (Naz et al., 2015). For example, microorganisms can change heavy metals from one oxidative state or organic complex to another (Xiong et al., 2010). Primarily, microorganisms put together for remediation depend with respect to the obstruction of the used organism to the substantial metal that is either actuated freely or through metal pressure (Naz et al., 2015). Regularly, microorganisms work through biosorption, bioaccumulation or bio-surfactant creation. Biosorption is a one of a kind remediation strategy where by microorganisms tie to and concentrate contaminations from the environment (Joutey et al., 2015). It is a cost-effective strategy for natural remediation from agricultural and industrial pollution sources (Fomina and Gadd, 2014; Javanbakht et al., 2014). A few microorganisms have been seen to have potential as biosorbents, for example, *Magnetospirillum gryphiswaldense* and *Bacillus subtilis*; fungi, for example, yeast, *Rhizopus arrhizus* and *Saccharomyces cerevisiae*; and algae, for example, *Chaetomorpha linum* (Romera et al., 2006; Wang and Chen, 2008; Zhou et al., 2012).

The presentation of genetic engineering of bacteria and different microorganisms opened up new skylines for planning custom-made organisms with engineered metal-restricting peptides that have improved affinity and selectivity for target metals to be utilized as heavy metal biosorbents. In bacteria, the surface presentation frameworks in different species, for example, *Staphylococcus xylosus* and *S. carnosus*, have been demonstrated to communicate diverse polyhistidyl peptides, that is, His3-Glu-His3 and His6. The encoded gene results of the chimeric surface proteins were focused to improve usefulness in terms of metal binding as surface openness (Samuelson et al., 2000).

The significant roles of novel genes that are effective in the bioaccumulation procedure and reactions to heavy metals in all organisms simultaneously has been outlined utilizing transcriptomics investigation (Leung et al., 2014; Shi et al., 2015). Besides, bioinformatics and mathematical modeling have been used to research the properties and possibilities of up-and-comer organisms so as to anticipate the grouping of synthetic compounds that can be endured by them (Stadnicka et al., 2012).

The process of utilizing plants in tidying up of contaminated soil or lethal environmental synthetic compounds and metalloids is known as phytoremediation (Tanee et al., 2009). This procedure has not gotten a lot of stress because of the low impacts of heavy metal on the phytoremediating plant (Dhankher et al., 2011; Musa and Ikhajiagbe, 2019). Right now, that are normally hyper-accumulators of heavy metals are utilized, yet even as that, such plants need biomass required for powerful remediation because of the measure of contaminations present and the geochemistry of the soil.

For this reason, biotechnologists have improved the gene of such plants to upgrade plant resistance and bioaccumulation potential through genetic engineering methodologies. Moreover, ongoing investigations, for example, Dhankher et al. (2011), have utilized OMICS innovations, for example, genomics, proteomics and transcriptomics, to explain the genetic determinants and procedures in heavy metals resilience and gathering in plants. These strategies have effectively been utilized in the phytoremediation of mercury (Hg), cadmium (Cd), lead (Pb), copper (Cu) and arsenic (As) (Figure 4.1).

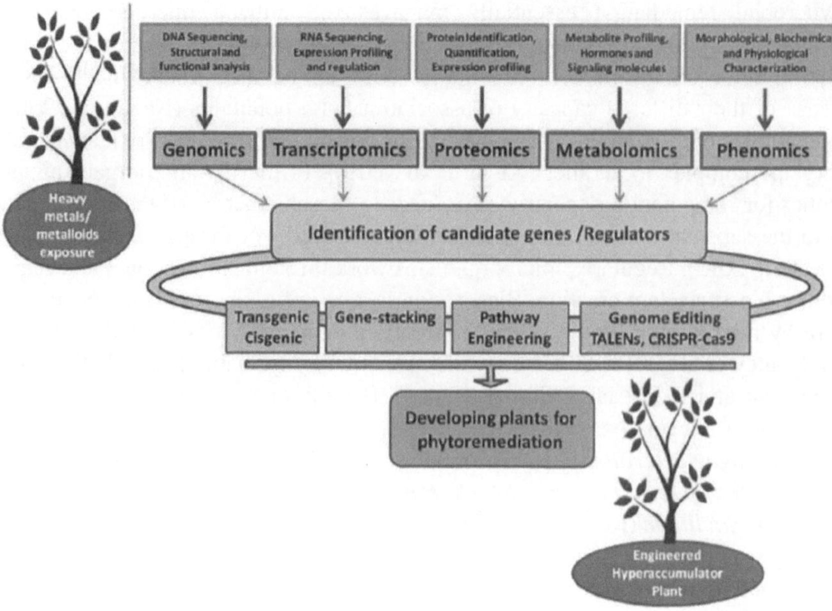

FIGURE 4.1 Biotechnological improvements in phytoremediation (Asubiaro et al., 2018).

Likewise, ongoing biotechnology has controlled pollutant transporter gene and take-up systems to improve remediation and resistance. For instance, Song et al. (2003) found that *Arabidopsis thaliana* that overexpresses yeast YCF1 (YeastCadmiumFactorl) showed more resistance and gathered more elevated level of Pb(II) and Cd(II). YCF1 is associated with Cd transporting to vacuoles by conjugation with glutathione (GSH). Overexpression of full-length *Nicotiana tabacum* plasma membrane channel protein (NtCBP4) indicated Pb2C hyperaccumulative potential (Sunkar et al., 2000). Hirschi et al. (2000) saw that tobacco plants can express CAX2 (calcium exchanger^2) which helps in hyperaccumulation of Ca^{2+}, Cd^{2+} and Mn^{2+} and furthermore improves plant resilience to Mn^{2+}. Progression in biotechnological approaches has additionally prompted the utilization of genome-wide expression analysis of plants roots, for example, rice that is exposed to different natural pollutants, for example, Cd, Cr(VI) and Pb (Dubey et al., 2014).

Metallic and Metalloid pollutant ligands have additionally been upgraded as a feature of biotechnological headway in phytoremediation. The overexpression of AtPCS1 in *A. thaliana* prompted improved tolerance to arsenate (Li et al., 2004). Additionally, *A. thaliana* plant with overexpression of aMTgene, PsMTA from pea (*Pisum sativum*), demonstrated expanded Cu^{2+} accumulation in roots (Evans et al., 1992). In addition, overexpression of AtPCS1 in *B. juncea* demonstrated a normal tolerance to soil arsenate (Gasic and Korban, 2007).

Biotechnologists have likewise enhanced the transformation of harmful metals and metalloids into less lethal synthetic compounds utilizing plants as a technique of environmental remediation. Transgenic plants that can improve selenium volatilization rates have been created (LeDuc et al., 2004).

Improvement in bioremediation technology may also involve enhancing capacity for production of metabolites and degradative enzymes by soil biological sentinels like earthworms. Vermiremediation is the utilization of earthworms to tidy up contaminants from the soil (Asubiaro et al., 2018). Studies on the possible utilization of animals, for example, earthworms in remediation of various pollutants have yielded a positive outcome (Gifford et al., 2006). Njoku et al. (2016) have seen that earthworms are tolerant to, and can expel, or help the expulsion of a wide scope of organic and inorganic contaminants, for example, pesticides, PAHs, crude oil, and heavy metals from the soil. It can likewise improve the remediation activities of the microbial community of the soil with the assistance of a novel microorganism got from the earthworm activities. A few scientists prescribed the potential ways to deal with vermiremediation which incorporates direct use of earthworms to contaminated soils and co-use of earthworms to polluted soils (Hickman, 2008; Dada, 2015).

Vermiremediation is financially savy and environmental maintainable method for mitigating natural pollutants, and it takes just few months (Njoku et al., 2016). The two significant tests in vermiremediation are earthworm survival and mobility (Sinha et al., 2008). In view of these, a few biotechnological upgrades have been acquainted to improve heavy metal remediation by earthworms (Gerhardt et al., 2017; Hussain et al., 2018), for example, expanded contaminant mobility and bioavailability with the assistance of biosurfactants and expanded overall environment worm growth with the assistance of a favorable situation and giving nourishment and ideal inoculum conditions, for example, pH, temperature, dampness and nourishment quality (Chirakkara et al., 2016). This assists with increasing the effectiveness and take-up time of the earthworm involved. Earthworms have demonstrated a dietary preference, which shows that their growth may be improved by improving the nourishment quality (Curry and Schmidt, 2007). Increased earthworm development prompts increased rates of take-up and natural contaminant degradation in the soil due to drilosphere impact.

Recently, specialists have utilized microbe-earthworm interactions in improving remediation utilizing earthworms. Nneji et al. (2016) observed that combined treatment with *A. millisoni* and *Bacillus* spp. brought about a huge decrease altogether in petroleum hydrocarbons and expanded absolute natural carbon and N in the soil compared with different medicines. Sun et al. (2011) announced that the level of pyrene expulsion within the sight of *E. foetida* was higher than in the treatment without earthworms at 14 days. Li et al. (2015) reported that the vaccination with both earthworms' species *E. fetida* and *A. robustus* altogether improved soil pentachlorophenol (PCP) vanishing and basal breath. Agronomic practices and soil amendments have additionally been utilized to delicate the soil so as to give the earthworms opportunities to burrow and improve the soil condition (Gerhardt et al., 2017).

4.13 CONCLUSION

Despite the fact that bioremediation is a successful method in the recovery of contaminated environment, there are still difficulties that should be looked into during the procedure of recovery. The difficulties are not far-fetched. There is a lack of specialized skill and information helps in understanding the mechanisms engaged with bioremediation. Remediation of some polluted sites, particularly those contaminated by normal mining, could be cost concentrated and troubled with environmental effects which makes sourcing for funds difficult.

In treatment of waterways and wastewater, the bioremediators may wind up contaminating the site because of copious supplements. Microorganisms may now and then be physiologically unfit, in any event, when environmental conditions are ideal, of lessening contamination focus to lower, health-based levels because of the take-up and digestion of the organic compounds halting at low concentration (Speight, 2018). If the lethality of the pollutant is high, it will affect the organism and render it useless. All these are problems begging for solution especially with the application of biotechnology. The fact that the pollutant might be nonbiodegradable is also a problem. Therefore it is important to look inwards. Undegradable contaminants can be contained, and as such with the turn of biotechnology, plant capacities as well as capacities or useful bio-entities can be enhanced to have capacities for containing or sequestering these undegradables.

Although the environmental conditions may not be suitable for the organism required for the tidy up processes and the environmental conditions may likewise meddle with the application procedure, a thorough study of environmental processes as well as an insight into gene-environment interactions can throw more light in ways to overcome these seemingly shortcomings of bioremediation.

REFERENCES

Asubiaro, D. O., Njoku, K. L., Amechina, N. S. and Musa, S. I. (2018). Enumeration and identification of microbial load in oil spilled soil remediated using *Eudrilus eugeniae*. Journal of Natural Products and Resources 4(1): 171–173.

Chen, L., Chen, W., Chen, L. and Chen, Y. (2009). Study on fungi-bacteria augmented remediation of petroleum contaminated soil from northwest of China. *Journal of Food, Agriculture and Environment* 7: 3–4. pt. 2: 750–753.

Chirakkara, R. A., Cameselle, C. and Reddy, K. R. (2016). Assessing the applicability of phytoremediation of soils with mixed organic and heavy metal contaminants. *Reviews in Environmental Science and Bio/Technology* 15: 299–326.

Christin, A. and Glenn, J. (2016). Mycoremediation of petroleum: a literature review. *Journal of Environmental Science and Engineering A* 5: 397–405.

Coelho, L. M., Rezende, H. C., Coelho, L. M., Sousa, P. A. R., Melo, D. F. O. and Coelho, N. M. M. (2015). Bioremediation of polluted waters using microorganisms. In Naofumi, S. (Ed.), *Advances in Bioremediation of Wastewater and Polluted Soil* (1st ed., pp. 1–22), InTech, Croatia.

Curry, J. P. and Schmidt, O. (2007). The feeding ecology of environment worms – a review. *Pedobiologia* 50: 463–477.

Cvjetko, P., Zovko, M. and Balen, B. (2014). Proteomics of heavy metal toxicity in plants. *Archives of Industrial Hygiene and Toxicology* 65: 1–18.

Dada, E. (2015). *Heavy Metal Remediation Potential of a Tropical Wetland Environment worm Species, Libyodrilus Violaceus*, Ph.D. Thesis, Department of Cell Biology and Genetics, University of Lagos, Lagos, Nigeria.

Das, S. and Dash, H. R. (2014). Microbial bioremediation: a potential tool for restoration of contaminated areas. In Surajit, D. (Ed.), *Microbial Biodegradation and Bioremediation* (1st ed., pp. 1–21), Elsevier Science, New York.

Del Giudice, L., Massardo, D. R., Pontieri, P., Bertea, C. M., Mombello, D. and Carata, E. (2008). The microbial community of Vetiver root and its involvement into essential oil biogenesis. *Environmental Microbiology* 10: 2824–2841.

Dhankher, O. P., Pilon-Smits, E. A. H., Meagher, R. B. and Doty, S. (2011). Biotechnological approaches for phytoremediation. In Altmanand, A. and Hasegawa, P. M. (Eds.), *Plant Biotechnology and Agriculture* (pp. 309–328), Academic Press, Oxford.

Dubey, S., Shri, M., Misra, P., Lakhwani, D., Bag, S. K. and Asif, M. H. (2014). Heavy metals induce oxidative stress and genome-wide modulation in transcriptome of rice root. *Functional and Integrative Genomics* 14: 401–417.

Elisashavili, V., Kachlishvili, E. and Penninckx, M. J. (2008). Lignocellulolytic enzymes profile during growth and fruiting of *Pleurotus ostreatus* on wheat straw and tree leaves. *Acta Microbiologica et Immunologica Hungarica* 55(2): 157–168.

Elmer, P. A. G. and Reglinski, T. (2006). Biosuppression of *Botrytis cinerea* in grapes. *Plant Pathology* 55: 155–177.

Evans, K. M., Gatehouse, J. A., Lindsay, W. P., Shi, J., Tommey, A. M. and Robinson, N. J. (1992). Expression of the pea metallothione in-like gene PsMTA in *Escherichia coli* and *Arabidopsis thaliana* and analysis of trace metal ion accumulation: implications for PsMTA function. *Plant Molecular Biology* 20: 1019–1028.

Felix, M. O., Charlotte, C. N. and Saheed, I. M. (2020). Bioremediation of diesel polluted soils with Eichhornia crassipes (water hyacinth). Notulae Scientia Biologicae 12(4): 920–928, DOI: 10.15835/12410814.

Fillion, M. (2008). Do transgenic plants affect rhizobacteria? *Microbial Biotechnology* 1: 463–475.

Finnerty, W. (1994). Biosurfactants in environmental biotechnology. *Current Opinion in Biotechnology* 5: 291–295.

Folch, A., Vilaplana, M., Amado, L., Vicent, R. and Caminal, G. (2013). Fungal permeable reactive barrier to remediate groundwater in an artificial aquifer. *Journal of Hazardous Materials* 262: 554–560.

Fomina, M. and Gadd, G. M. (2014). Biosorption: current perspectives on concept, definition and application. *Bioresources Technology* 160: 3–14.

Frascari, D., Zanaroli, G. and Danko, A. S. (2015). *In situ* aerobic cometabolism of chlorinated solvents: a review. *Journal of Hazardous Materials* 283:382–399.

Frutos, F. J. G., Pe´rez, R., Escolano, O., Rubio, A., Gimeno, A., Fernandez, M. D., Carbonell, G., Perucha, C. and Laguna, J. (2012). Remediation trials for hydrocarbon-contaminated sludge from a soil washing process: Evaluation of bioremediation technologies. *Journal of Hazardous Materials* 199: 262–271.

Garbisu, C. and Alkorta, I. (2003). Basic concepts on heavy metal soil bioremediation. *European Journal of Mineral Processing and Environmental Protection* 3(1): 58–66.

Gasic, K. and Korban, S. S. (2007). Transgenic Indian mustard (*Brassica juncea*) plants expressing an Arabidopsis phytochelatin synthase (AtPCS1) exhibit enhanced As and Cd tolerance. *Plant Molecular Biology* 64: 361–369.

Gerhardt, K. E., Gerwing, P. D. and Greenberg, B. M. (2017). Opinion: taking phytoremediation from proven technology to accepted practice. *Plant Science* 256: 170–185.

Gifford, S., Dunstan, R., O'Connor, W., Koller, C. and MacFarlane, G. (2006). Aquatic zooremediation: deploying animals to remediate contaminated aquatic environments. *Trends in Biotechnology* 25: 60–65.

Giraud, F., Guiraud, P., Kadri, M., Blake, G. and Steiman, R. (2001). Biodegradation of anthracene and fluoranthene by fungi isolated from an experimental constructed wetland for wastewater treatment. *Water Research* 35 (17): 4126–4146.

Gomez, F. and Sartaj, M. (2013). Field scale *ex situ* bioremediation of petroleum contaminated soil under cold climate conditions. *International Biodeterioration and Biodegradation* 85: 375–382.

Harms, H., Schlosser, D. and Wick, L. Y. (2011). Untapped potential: exploiting fungi in bioremediation of hazardous chemicals. *Nature Reviews: Microbiology* 9: 117–192.

He, H., Shen, B., Korshalla, J. and Carter, G. T. (2001). Circulocins, new antibacterial lipopeptides from *Bacillus circulans*, J2154. *Tetrahedron* 57: 1189–1195.

He, M., Li, X., Liu, H., Miller, S. J., Wang, G. and Rensing, C. (2011). Characterization and genomic analysis of a highly chromate resistant and reducing bacterial strain Lysinibacillus fusiformis ZC1. *Journal of Hazardous Materials* 185: 682–688.

Hickman, Z. and Reid, B. (2008). Environment worm assisted bioremediation of organic contaminants. *Environment International* 34: 1072–1081.

Hirschi, K. D., Korenkov, V. D., Wilganowski, N. L. and Wagner, G. J. (2000). Expression of Arabidopsis CAX2 in tobacco. Altered metal accumulation and increased manganese tolerance. *Plant Physiology* 124: 125–133.

Hossain, Z. and Komatsu, S. (2012). Contribution of proteomic studies towards understanding plant heavy metal stress response. *Frontiers in Plant Science* 3: 310.

Hu, P., Brodie, E. L., Suzuki, Y., McAdams, H. H. and Andersen, G. L. (2005). Whole-genome transcriptional analysis of heavy metal stresses in *Caulobacter crescentus. Journal of Bacteriology* 187: 8437–8449.

Hussain, I., Puschenreiter, M., Gerhard, S., Schöftner, P., Yousaf, S., Wang, A., Syed, J.H. and Reichenauer, T. G. (2018). Rhizoremediation of petroleum hydrocarbon-contaminated soils: improvement opportunities and field applications. *Environmental and Experimental Botany* 147: 202–219.

Igiebor, F. A., Osarumwense, J. O., Obinyan, B. O. and Okoye, P. C. (2017). Isolation and identification of indigenous hydrocarbon tolerant fungi from soil contaminated with Biodiesel in Benin City, Nigeria. *International Journal of Agriculture and Environmental Science* 4 (6): 47–50.

Ikhajiagbe, B, Omoregie, G. O. and Ekhator, P. O. (2017). Morphological adaptation of Vernonia Amygdalina in a cadmium-spiked soil influenced by periodic soil wetting with grey water. *FUPRE Journal of Scientific and Industrial Research* 1(2): 63–75.

Javanbakht, V., Alavi, S. A. and Zilouei, H. (2014). Mechanisms of heavy metal removal using microorganisms as biosorbent. *Water Science and Technology* 69: 1775–1787.

Jobby, R. and Desai, N. (2017). Bioremediation of heavy metals. In Pravinder, K., Bhola, R. G., and Govil, J. N. (Eds.), *Environmental Science and Engineering: Biodegradation and Bioremediation* (1st ed., pp. 201–220), Studium Press LLC.

Joutey, N. T., Sayel, H., Bahafid, W. and ElGhachtouli, N. (2015). Mechanisms of hexavalent chromium resistance and removal by microorganisms. *Reviews of Environmental Contamination and Toxicology* 233: 45–69.

Khudur, L. S., Shahsavari, E., Miranda, A. F., Morrison, P. D., DayanthiNugegoda, D. and Ball, A. S. (2015). Evaluating the efficacy of bioremediating a diesel-contaminated soil using ecotoxicological and bacterial community indices. *Environmental Science and Pollution Research* 22 (14809): 14819.

Kim, S., Krajmalnik-Brown, R., Kim, J. O. and Chung, J. (2014). Remediation of petroleum hydrocarbon-contaminated sites by DNA diagnosis-based bioslurping technology. *Science of the Total Environment* 497: 250–259.

Kohlmeier, S., Smits, T. H. M., Ford, R. M., Keel, C., Harms, H. and Wick, L. Y. (2005). Taking the fungal highway: mobilization of pollutant-degrading bacteria by fungi. *Environmental Science and Technology* 39: 4640–4646.

Kotrba, P., Najmanova, J., Macek, T., Ruml, T. and Mackova, M. (2009). Genetically modified plants in phytoremediation of heavy metal and metalloid soil and sediment pollution. *Biotechnology Advances* 27: 799–810.

Krzmarzick, M. J, Taylor, D. K., Fu, X. and McCutchan, A. L. (2018). Diversity and niche of archaea in bioremediation. *Archaea (Vancouver, B.C.)* 12: 1–17.

LeDuc, D. L., Tarun, A. S., Montes-Bayon, M., Meija, J., Malit, M. F. and Wu, C. P. (2004). Overexpression of selenocyste in emethyl transferase in Arabidopsis and Indian mustard increases selenium tolerance and accumulation. *Plant Physiology* 135: 377–383.

Leonardi, V., Šašek, V., Petruccioli, M., D'Annibale, A., Erbanová, P. and Cajthaml, T. 2006. Bioavailability modification and fungal biodegredation of PAHs in aged industrial soils. *International Biodeterioration and Biodegradation* 60: 165–170.

Leung, P. T. Y., Ip, J. C. H., Mak, S. S. T., Qiu, J. W., Lam, P. K. S. and Wong, C. K.C. (2014). De novo transcriptome analysis of *Pernaviridis* highlights tissue specific patterns for environmental studies. *BMC Genomics* 15: 804.

Li, X., Lin, Z., Luo, C., Bai, J., Sun, Y. and Li, Y. (2015). Enhanced microbial degradation of pentachlorophenol from soil in the presence of environment worms: evidence of functional bacteria using DNA-stable isotope probing. *Soil Biology and Biochemistry* 81: 168–177.

Li, Y., Dhankher, O. P., Carreira, L., Lee, D., Chen, A. and Schroeder, J. I. (2004). Overexpression of phytochelatin synthase in Arabidopsis leads to enhanced arsenic tolerance and cadmium hypersensitivity. *Plant Cell Physiology* 45: 1787–1797.

Li, Y. Q., Liu, H. F., Tian, Z. L., Zhu, L. H., Wu, Y. H. and Tang, H. Q. (2008). Diesel pollution biodegradation: Synergetic effect of mycobacterium and filamentous fungi. *Biomedical and Environmental Sciences* 21 (3): 181–187.

Madhavi, V. and Lele, S. S. (2009). Laccase: properties and applications. *Bioresources* 4 (4): 1694–1717.

Meulenberg, R., Rijnaarts, H. H., Doddema, H. J. and Field, J. A. (1997). Partially oxidized polycyclic aromatic hydrocarbons show an increased bioavailability and biodegradability. *FEMS Microbiology Letters,* 152(1): 45–49.

Meysami, P., and Baheri, H. (2003). Pre-screening of fungi and bulking agents for contaminated soil bioremediation. *Advances in Environmental Research* 7 (4): 881–887.

Morita, T., Konishi, M., Fukuoka, T., Imura, T. and Kitamoto, D. (2007). Microbial conversion of glycerol into glycolipid biosurfactants, mannosylerythritol lipids, by a basidiomycete yeast, *Pseudozyma antarctica* JCM 10317T. *Journal of Bioscience and Bioengineering* 104 (1), 78–81.

Muralidharan, S., Thompson, E., Raftos, D., Birch, G. and Haynes, P. A. (2012). Quantitative proteomics of heavy metal stress responses in Sydney rock oysters. *Proteomics* 12: 906–921.

Musa, S. I. (2019). Isolation and identification of diesel oil-degrading bacteria in used engine oil contaminated soil. Journal of Applied Science and Environmental Management 23(3): 431–435.

Musa, S. I., Foluke, A., Njoku, K. L. and Ndiribe, C. C. (2018). The effect of bioremediated diesel polluted soil ammended with *Eicchorniacrassipes* (Water Hyacinth) on the germination parameters of Amaranthushybridus L. (Green Amaranth). *International Journal of Science and Research* 7(1): 1686–1688.

Musa, S. I., Awayewaserere, K. O., Njoku, K. L. (2019). Effects of Dump Site Soil on the Leaf Structures of Luffa cylindrical (Sponge gourd) and Amaranthus viridis (Green Amaranth). Journal of Applied Science and Environmental Management 23(2): 307–311.

Musa, S. I. and Ikhajiagbe, B. (2019). Bioaccumulation of heavy metals by *Vignaunguiculata* (cowpea) grown on Olusosun Dump Site soil, Lagos, Nigeria. *Journal of Underutilized Legumes* 1(1): 91–98.

Naz, T., Khan, M. D., Ahmed, I., Rehman, S. U., Rha, E. S. and Malook, I. (2015). Biosorption of heavy metals by Pseudomonas species isolated from sugar industry. *Toxicology and Industrial Health* 32(9): 1619–1627.

Njoku, K., Akinola, M. and Anigbogu, C. (2016). Vermiremediation of soils contaminated with petroleum products using Eiseniafetida, *Journal of Applied Sciences and Environmental Management* 20: 771–779.

Nneji, L. M., Somade, O. T. and Adeyi, A. O. (2016). Environment worm-assisted bioremediation of petroleum hydrocarbon–contaminated soils from motorcar mechanic workshops in Ibadan, Oyo State, southwestern Nigeria. *Bioremediation Journal* 20: 263–285.

Novotný, Č., Svobodová, K., Erbanová, P., Cajthaml, T., Kasinath, A., Lang, E. and Šašek, V. (2004). Ligninolytic fungi in bioremediation: extracellular enzyme production and degradation rate. *Soil Biology and Biochemistry* 36: 1545–1551.

Nwachukwu, A. M. (2014). Prospective techniques for in-situ treatment and protection of aquifers: A sustainable hydrology review. *International Journal of Water Resources and Environmental Engineering* 6(4): 131–143.

Oberbremer, A., Müller-Hurtig, R. and Wagner, F. (1990). Effect of the addition of microbial surfactants on degradation in a soil population in a stirred reactor. *Applied Microbiology and Biotechnology* 32: 485–489.

Okparanma, R. N., Ayotamuno, J. M., Davis, D. D. and Allagoa, M. (2011). Mycoremediation of polycyclic aromatic hydrocarbons (PAH)-contaminated oil-based drill-cuttings. *African Journal of Biotechnology* 10 (26): 5149–5156.

Osarumwense, J. O. and Igiebor F. A. (2018). Assessment of indigenous bacteria from biodiesel effluents contaminated site. *Journal of Applied Sciences and Environmental Management* 22 (2): 157–160.

Pathak, V. M. and Navneet. (2017). Review on the current status of polymer degradation: a microbial approach. *Bioresources and Bioprocessing* 4(1): 1–31.

Peña-Montenegro, T. D. and Dussán, J. (2013). Genome sequence and description of the heavy metal tolerant bacterium *Lysinibacillussphaericus* strain OT4b.31.Stand. *Genomic Science* 9: 42–56.

Philp, J. C. and Atlas, R. M. (2005) Bioremediation of contaminated soils and aquifers. In: Atlas, R. M., Philp. J. C. (eds) *Bioremediation: Applied Microbial Solutions for Real-World Environmental Clean-Up.* (pp 139–236), American Society for Microbiology (ASM) Press, Washington, DC.

Pinedo-Rivilla, C., Aleu, J. and Collado, I. G. (2009). Pollutants biodegradation by fungi. *Current Organic Chemistry* 13: 1194–1214.

Pointing, S. B. (2001). Feasibility of bioremediation by white-rot fungi. *Applied Microbiology and Biotechnology* 57: 20–33.

Pozdnyakova, N. (2012). Involvement of the ligninolytic system of white-rot and litter- decomposing fungi in the degradation of polycyclic aromatic hydrocarbons. *Biotechnology Research International* 3: 1–20.

Prenafeta-Boldu, F. X., Kuhn, A., Luykx, D. M. A. M., Anke, H., van Groenestijn, J. W. and de Bont, J. A. M. (2001). Isolation and characterisation of fungi growing on volatile aromatics hydrocarbons as their sole carbon and energy source. *Mycological Research* 105 (4): 477–484.

Pushpanathan, M., Jayashree, S., Gunasekaran, P. and Rajendhran, J. (2014). Microbial bioremediation: a metagenomic approach. In Surajit, D. (Ed.), *Microbial Biodegradation and Bioremediation* (1st ed., pp. 407–419), Elsevier Science, New York.

Reddi, L. N. and Inyang, H. I. (2000). *Geoenvironmental Engineering: Principles and Applications.* (1st ed.). Marcel Dekker, New York.

Robert, M., Mercade, M. E., Bosch, M. P., Parra, J. L., Espuny, M. J., Manresa, M. A. and Guinea, J. (1989) Effect of the carbon source on biosurfactant production by *Pseudomonas aeruginosa* 44Ti. *Biotechnology Letters* 11: 871–874.

Romera, E., González, F., Ballester, A., Blázquez, M. L. and Muñoz, J. A. (2006). Biosorption with algae: a statistical review. *Critical Reviews in Biotechnology* 26: 223–235.

Roy, M., Giri, A. K., Dutta, S. and Mukherjee, P. (2015). Integrated phytobial remediation for sustainable management of arsenic in soil and water. *Environment International* 75: 180–198.

Russell, J. R., Huang, J., Anand, P., Kucera, K., Sandoval, A. G., Dantzler, K. W., et al. (2011). Biodegradation of polyester polyurethane by endophytic fungi. *Applied and Environmental Microbiology* 77 (17): 6076.

Ryan, D., Sinha, R. K. and Bharambe, G. (2008). Converting wasteland into wonderland by environment worms a low-cost nature's technology for soil remediation: a case study of vermiremediation of PAHs contaminated soil. *The Environmentalist* 28: 466–475.

Samuelson, P., Wernérus, H., Svedberg, M. and Ståhl, S. (2000). Staphylococcal surface display of metal-binding polyhistidyl peptides. *Applied Environmental Microbiology* 66: 1243–1248.

Shi, B., Huang, Z., Xiang, X., Huang, M., Wang, W. X. and Ke, C. (2015). Transcriptome analysis of the key role of GAT2 gene in the hyper-accumulation of copper in the oyster *Crassostrea angulata*. *Scientific Reports* 5: 17751.

Singh, B. K., Millard, P., Whiteley, A. S. and Murrell, J. C. (2004) Unravelling rhizosphere-microbial interactions: opportunities and limitations. *Trends in Microbiology* 12: 386–393.

Sinha, R. K., Bharambe, G. and Ryan, D. (2008). Converting wasteland into wonderland by environment worms a low-cost nature's technology for soil remediation: a case study of vermiremediation of PAHs contaminated soil. *The Environmentalist* 28: 466–475.

Singh, R. P., Dhania, G., Sharma, P. A. and Jaiwal, K. (2006). Biotechnological approaches to improve phytoremediation efficiency for environment contaminants. In Singh, S. N., and Tripathi, R. D. (Eds.), *Environmental Bioremediation Technologies* (pp. 223–258), Springer, New York.

Smith, E., Thavamani, P., Ramadass, K, Naidu, R., Srivastava, P. and Megharaj, M. (2015). Remediation trials for hydrocarbon-contaminated soils in arid environments: evaluation of bioslurry and biopiling techniques. *Integrative Biodeterioration and Biodegradation* 101: 56–65.

Somdee, T., Thathong, B. and Somdee, A. (2015). The removal of cyanobacterial hepatotoxin [Dha(7)]microcystin-LR via bioaccumulation in water lettuce (*Pistia stratiotes* L.). *Bulletin of Environmental Contamination and Toxicology* 96: 388–394.

Song, W. Y., Sohn, E. J., Martinoia, E., Lee, Y. J., Yang, Y. Y. and Jasinski, M. (2003). Engineering tolerance and accumulation of lead and cadmium in transgenic plants. *Nature Biotechnology* 21: 914–919.

Sorensen, J. and Sessitsch, A. (2007). Plant-associated bacteria– lifestyle and molecular interactions. In van Elsas, J. D., Jansson, J. D., and Trevors, J. T. (Eds.), *Modern Soil Microbiology* (2nd ed., pp. 211–236), CRC Press, Boca Raton, FL.

Speight, J. G. (2018). Biological transformations. In Bridget, H. (Ed.), *Reaction Mechanisms in Environmental Engineering: Analysis and Prediction* (1st ed., pp. 269–306.), Butterworth-Heinemann, Oxford.

Stadnicka, J., Schirmer, K. and Ashauer, R. (2012). Predicting concentrations of organic chemicals in fish by using toxicokinetic models. *Environmental Science and Technology* 46: 3273–3280.

Sun, H., Li, J., Wang, C., Wang, L. and Wang, Y. (2011). Enhanced microbial Removal of pyrene in soils in the presence of environment worms. *Soil and Sediment Contamination* 20: 617–630.

Sunkar, R., Kaplan, B., Bouché, N., Arazi, T., Dolev, D. and Talke, I. N. (2000). Expression of a truncated tobacco NtCBP4 channel in transgenic plants and disruption of the homologous Arabidopsis CNGC1 gene confer Pb_2C tolerance. *Plant Journal* 24: 533–542.

Tanee, F. and Akonye, L. (2009). Effectiveness of *Vigna unguiculata* as a phytoremediation plant in the remediation of crude oil polluted soil for cassava (*Manihot esculenta*; Crantz) cultivation. *Journal of Applied Science and Environmental Management* 13: 43–47.

Tychanowitcz, G. K., de Souza, D. F., Souza, C. G. M., Kadowaki, M. K. and Peralta, R. M. (2006). Copper improves the production of laccase by the white-rot fungus *Pleurotuspulmonarius* in solid state fermentation. *Brazilian Archives of Biology and Technology* 49 (5): 699–704.

Vance-Harrop, M. H., De Gushusãmao, N. B. and De Campos-Takaki, G. M. (2003). New bioemulsifiers produced by Candida lipolytica using D-Glucose and Babassu oil as carbon sources. *Brazilian Journal of Microbiology* 34: 120–123.

Verma, J. P. and Jaiswal, D. K. (2016). Book review: advances in biodegradation and bioremediation of industrial waste. *Frontiers in Microbiology* 6: 1–2.

Vogt, C. and Richnow H. H. (2013). Bioremdiation via *in situ* microbial degradation of organic pollutants. In Schippers A., Glombitza F., Sand W. (Eds.) *Geobiotechnology II. Advances in Biochemical Engineering/Biotechnology* (142, p), Springer Verlag, Berlin.

Wang, J. and Chen, C. (2008). Biosorbents for heavy metals removal and their future. *Biotechnology Advances* 27: 195–226.

Wick, L. Y., Furuno, S., and Harms, H. (2010). *Handbook of Hydrocarbon Microbiology* Edited by Timmis, K. N., McGenity, T., van der Meer, J. R. and de Lorenzo, V. Springer, New York.

Wong, M. H. and Chu, L. M. (1985). Yield and metal uptake of *Cynodondactylon* (Bermuda grass) grown in refuse-compost-amended soil. *Agriculture, Ecosystems and Environment* 14 (1–2): 41–52.

Wong, M. H. and Lau, W. M. (1985). Root growth of *Cynodon* and *Eleusineindica* collected from motorways at different concentrations of lead. *Environmental Research* 36: 257–267.

Wu, J. Y., Ng, J., Shi, M., and Wu, S. J. (2007). Enhanced secondary metabolite (tanshinone) production of Salvia miltiorrhiza hairy roots in a novel root-bacteria coculture process. *Applied Microbiology and Biotechnology* 77: 543–550.

Xiong, J., Wu, L., Tu, S., VanNostrand, J. D., He, Z. and Zhou, J. (2010). Microbial communities and functional genes associated with soil arsenic contamination and the rhizosphere of the arsenic-hyperaccumulating plant *Pteris vittata* L. *Applied and Environmental Microbiology* 76: 7277–7284.

Yan, Q.-X., Hong, Q., Han, P., Dong, X.-J., Shen, Y. J. and Li, S. P. (2007). Isolation and characterization of a carbofuran-degrading strain Novosphingobium sp. FND-3. *FEMS Microbiology Letters* 271: 207–213.

Young, D., Rice, J., Martin, R., Lindquist, E., Lipzen, A. and Grigoriev, I. (2015). Degradation of bunker C fuel oil by white-rot fungi in sawdust cultures suggests potential applications in bioremediation. *PLoS One* 10 (6): e0130381.

Zakeri, F., Sadeghizadeh, M., Kardan, M. R., ShahbaniZahiri, H., Ahmadian, G. and Masoumi, F. (2012). Differential proteome analysis of a selected bacterial strain isolated from a high background radiation are in response to radium stress. *Journal of Proteomics* 75: 4820–4832.

Zhang, R., Cui, Z., Jiang, J., Gu, X. and Li, S. (2005). Diversity of organophosphorus pesticides degrading bacteria in a polluted soil and conservation of their organophosphorus hydrolase genes. *Canadian Journal of Microbiology* 5: 337–343.

Zhang, X. H. (2005). Remediation techniques for soil and groundwater. In Y. Qian. (Ed.), *Journal of Point Sources of Pollution: Local Effects and It's Control* 2: 350–364.

Zhongli, C., Shunpeng, L. and Guoping, F. (2001). Isolation of methyl parathion-degrading strain m6 and cloning of the methyl parathion hydrolase gene. *Applied and Environmental Microbiology* 67(10): 4922–4925.

Zhou, D., Li, Y., Zhang, Y., Zhang, C., Li, X., Chen, Z., Huang, J., Li, X., Flores, G. and Kamon, M. (2014). Column test-based optimization of the permeable reactive barrier (PRB) technique for remediating groundwater contaminated by landfill leachates. *Journal of Contaminant Hydrology* 168: 1–16.

Zitte, L. F., Awi-Waadu, G. D. B. and John, A. U. (2012). Effect of oyster mushroom (*Pleurotus ostreatus*) mycelia on petroleum hydrocarbon contaminated substrate. *Journal of Agriculture and Social Research* 12 (2): 115–121.

5 Development of Bioenhancement Formulations for Bioremediation of Hydrocarbon-Impacted Environments

M. B. Yerima, A. O. Adeleye, M. E. Nkereuwem, and V. O. Onokebhagbe
Federal University Dutse

K. E. Nwagu
Alex Ekwueme Federal University

Rebecca Richard
Federal University Gashua

I. K. Adam
Federal University Lafia

A. F. Umar
Abubakar Tafawa Balewa University

A. A. Farouq
Usmanu Danfodiyo University

CONTENTS

DOI: 10.1201/9781003180289-5

5.1 INTRODUCTION

Many authors (Ibe and Ibe, 1984; Ijah and George, 2000; Atuanya, 1987) have done and reported numerous research studies on biodegradation of crude oil and the bioremediating potentials of hydrocarbonoclastic microorganisms in cleaning up hydrocarbon-related pollution in Nigeria. Crude oil exploitation and exploration have been going on continuously in the Niger Delta area of Nigeria leading to the pollution of terrestrial and aquatic ecosystems. Most especially, potable water sources have been dealt a huge blow owing to the shallowness of the aquifers present therein. It has been documented that a whopping number of oil spills recorded between 1976 and 1996 stands as 4,647, while 2,369, 70.04 cm^3, account for the quantity of crude oil spilled. Again, 390 cases of oil spill were recorded in 1998 alone suggesting the hydrocarbon pollution related abuse that the region has been subjected to over the years.

Serious ecological problems ranging from water and soil pollution are witnessed when crude oil spills inadvertently. Pollution of the ground water due to the influence of hydrocarbon-related pollutants is widespread in any environment where exploration and exploitation of crude oil take place. According to Ijah and George (2000), water meant for drinking becomes undrinkable when crude oil or petroleum products even at extremely low concentrations are present. The alarming state of hydrocarbon-related pollution of the aquifers in the Niger Delta area of Nigeria and indeed some parts of Nigerian cities has got to a level of rendering drinking water unfit for human consumption.

When hydrocarbon pollution occurs on water bodies, it forms a thick layer that prevents penetration of atmospheric oxygen and its subsequent dissolution. This results in the depletion of dissolved oxygen and a stress factor is introduced. Fish and other aquatic resources begin to die since they can only utilize dissolved oxygen. Oil pollution disrupts the food chain and hence the ecosystem. Oil pollution disrupts the economy (fishing, recreational uses of water such as angling, yachting and swimming). For example, insects that are indigenous to shorelines and birds that visit such shorelines to pick fishes are trapped and killed. Their wings become smothered and this affects their ability to fly. Their biological roles of pollination and dispersal of fruits are then incapacitated. These have great implications and consequences on loss of biodiversity and possible extinction. Other effects of oil pollution on Biota such as birds include reduction in reproduction rate due to reduced egg production.

However, the advent of bioremediation has since made the task of remediating and reclaiming polluted ecosystems feasible. Attainment of bioremediation can be achieved through the adoption of biostimulation which is a strategy that involves the stimulation of the growth of indigenous oil degraders through enriching with limiting nutrients or other growth limiting co-substrates. According to many authors (Swannell et al., 1996; Wilfred et al., 2002; Ibekwe et al., 2006), this strategy is advantageous due to the fact that the employable hydrocarbonoclastic microorganisms have already adapted to the prevailing environmental factors present in their indigenous environments and the nutrients added as biostimulants could be utilized efficiently leading to optimum microbial growth and activity. Odokuma and Dickson (2003) reported that the objective of biostimulatory approach is to optimize the indigenous microorganisms in bringing about desirable degradation of the pollutants.

It has been documented that biostimulation can be attained through the addition of numerous forms of nutrients (phosphorus, nitrogen and other trace elements) that are limiting. It is worthwhile to reiterate that the delivery of nutrients should be done in a manner that allows the additives to be readily available for optimum microbial utilization. Yongabi et al. (2006) reportedly submitted that biostimulation normally leads to a substantial increase in heterotrophic and hydrocarbonoclastic bacterial populations due to the effect caused by supplementation with desirable nutrients. These authors concluded that addition of desirable nutrients to soil contaminated with enhanced biodegradation processes significantly in their study. Interestingly, Anonymous (1994) submitted that biostimulation is adjudged to be comparatively cheaper in lieu of using and depending on pure cultures for bioremediation projects. In this current study, sewage sludge that has been implicated as possessing mineralized product and other biostimulants that contains appreciable quantities of micronutrients were adopted with a view to testing their bioremediating capacity.

5.2 MATERIALS AND METHODS

5.2.1 Samples Collection and Processing

5.2.1.1 Collection of Sewage Sludge

Sewage sludge samples were collected and subjected to sun drying for a week, milled into powder with a grinding machine and subsequently kept in sterile dry polythene bags. Specifically, trace metals and phosphorus contents of the sewage sludge were analyzed using standard procedures.

5.2.1.2 Collection of Cow Dung and Cocoa Pod Husks (CPH)

About 30 kg cow dung (CD) used was collected from abattoir and dried using sun. The sample was then ground and sieved into fine powder. On the other hand, collection of cocoa pod husks (CPH) was done in a typical cocoa farm in Ekiti State, Nigeria. About 15 kg out of the collected CPH was subjected to air drying for the period of 30 days. It was subsequently milled into powder through the aid of an

FIGURE 5.1 CPH heaped up in a typical cocoa plantation.

electric blender as described by Agbor et al. (2015). The powdered material was sieved into fine powder and subsequently stored in containers and labeled accordingly. See below how abundant CPH is from a typical farm (Figure 5.1).

5.2.1.3 Composting of Biostimulated Agents

Compost was generated from the composting of 10 kg of fresh CPH and 10 kg of CD. This was done according to the procedure reported by Adegunloye and Olorunnusi (2016). Composting was done as depicted in Figure 5.2. It should be noted that sewage sludge was anaerobically digested.

5.2.1.4 Mineral Salt Medium Preparation

A number of salts (phosphorus, magnesium, sodium, ammonium and potassium salts) that have got vital nutrients needed by the targeted bacteria for growth were prepared according to the procedure outlined by Mukred et al. (2008).

5.2.1.5 Isolation of Spent Engine Oil Degrading Bacteria

The bacterial isolates that were used as coculture inoculants in this study were isolated from a spent engine oil polluted soil adopting the procedure of Adeleye and Yerima (2019).

5.2.1.6 Biodegradation Experiments

Thirty-six polyethylene bags were set up for each of the three biodegradation experiments conducted in this study. According to Nkereuwem et al. (2010) and Ezekoye

The mixture was put in a self-designed compost container and labeled "composted CPH + Cowdung 1:1 w/w." The mixture was continuously mixed as done by Vishan *et al.* (2017) for 20 days. Physicochemical properties were determined after curation.

FIGURE 5.2 Composting of CPH and cow dung in a self-designed compost container.

et al. (2017), apart from the 12 experimental bags that were used as controls in each of the experiments, 150 g of each sterilized biostimulant was added and thoroughly mixed with the soil contaminated with varying levels (5%, 10% and 15%) of SEO. Subsequently, bacterial cocultures (150 mL) was added. All the experimental bags were subsequently incubated at ambient temperature (36°C) for 70 days as outlined by Chorom et al. (2010). As established by Ayotamuno et al. (2006) and Chorom et al. (2010), the contents of each bag were tilled twice a week for aeration. Maintenance of moisture content was equally done twice a week by the addition of 6 mL sterile distilled water as done by Abioye et al. (2012). Periodic sampling for the estimations of Total Petroleum Hydrocarbons (TPH) and Polycyclic Aromatic Hydrocarbons (PAHs) was done at the commencement, fifth week and tenth week of the experiments.

5.2.1.7 Determination of Total Petroleum Hydrocarbon

TPH was determined using the USEPA 1850C method described by USEPA (2003) protocols. The cultures used in the degradation experiment conducted in this study were maintained in sterile minimal medium that was overlaid with crude oil. Already sterilized nutrient broth (10 mL) was dispensed into universal bottles and the test bacteria were subsequently inoculated into these media. The bottles were then incubated at 24°C for 48 hours. Eight milliliters of the broth culture that had the test bacteria actively growing was inoculated into another sets of universal bottles containing minimal medium that was overlaid with Bonny light crude oil. Appropriate labeling of the bottles with the respective concentrations of sewage sludge adopted was done. These bottles were subsequently left on the bench for 16 days. Gas chromatographic analysis of the un-degraded and residual crude from the universal bottles was done with a view to determining the TPH present in the samples.

5.2.1.8 Determination of Polycyclic Aromatic Hydrocarbons (PAHs)

PAHs were determined using the USEPA 1850C method described by USEPA (2003) protocols. The dichloromethane employed as an extractant was purchased from

Fischer Scientific (UK). A mixture of PAHs consisting of acenaphthene, benz (a) anthracene, fluoranthene benzo (a) pyrene, anthracene, naphthalene, fluorine, pyrene, chrysene, benzo (b) fluoranthene, benzo (k) fluoranthene, dibenzo (a, h) anthracene, acenaphthylene, benzo (ghi) perylene, indeno (1, 2, 2 d) pyrene and phenanthrene were prepared and utilized as standard. Additionally, internal standards containing four isotopically labeled PAHs, chrysene-$_{d12}$, penanthrene-$_{d10}$, acenaphthelene-$_{d10}$ and perylene-$_{d12}$ purchased from ChemService, Westchester, PA, were employed.

5.2.1.8.1 Procedure for Preparing Standard Solutions

The five standard solutions in which each contained 16 compounds prepared in the study were done through the dilution of the standard mix (i.e. mixture from NIST) to concentrations that were desirable with HPLC grade dichloromethane. Subsequently, 0.5 μg of internal standard was dispensed into the solutions. Capped and sealed vial was employed to store the final mixture until its deployment for analysis.

5.2.1.8.2 Extraction of Residual Oil from Degradation Bottles

The standard procedures established by USEPA and reported by Thomas (1990) were adopted for the extraction of residual oil from the degradation bottles. This method was staged by mounting a 100 mL capacity separating funnel on a retort stand. A muffle furnace operated at 150°C overnight was employed to sterilize the separating funnel. Dichloromethane was used to rinse the separating funnel for several minutes before it was used. Having rinsed the funnel, it was taken out and permitted to drain out totally in a fume cupboard. Addition of water-oil mixture (20 mL) derived from the degradation bottles to the funnel and dichloromethane (20 mL) was subsequently done. The mixture derived from these was vigorously shaken for 4 minutes and was left to set leading to its visible separation. After 10 minutes, the organic layer generated was effected. Subsequently, through the employment of the aqueous layer, the aforementioned procedure was made to run two times. Using a rotary evaporator, the three layers of the organic phase generated were mixed together and later allowed to evaporate to 1 mL. The recovery in terms of percentage was calculated using the mathematical formular depicted below:

$$\%r = q_d / q_a \times 100$$

where:

q_d = the quantity that was determined through analysis and
q_a = the quantity that was added

Calibration

Several dilutions that were derived from the standard PAHs mixtures were analyzed so as to estimate the limits of detection and quantification. The employment of triplicate analyses was adopted for such estimations.

5.2.1.8.3 Employment of Gas Chromatography and Mass Spectrophotometer (GC-MS) in Determining PAHs

An instrument known as Finnigan Magnum that was equipped with a CTC A200S autosampler and a 30 μm, 0.25 ID DB-5 MS fused silica capillary column was employed to run the GC-MS analysis. A carrier gas (Helium) and a column head pressure kept at 10 psi so as to obtain an expected flow rate of 1 mL/minute were adopted. The injector and transfer line were retained at 290°C and 250°C accordingly. All the injection volumes were kept at 1 µL in the split less mode. At 70°C for 4 minutes, the column temperature was at first maintained and was subsequently increased at the rate of 10°C per minute to 300°C. The temperature was further kept for another 10 minutes. The mass spectrophotometer (MS) was later employed in the detection of electron ionization, and all the spectra that emerged were derived through an automatic gain control and the employment of a mass range (m/z 50–400).

5.2.1.8.4 Identification and Quantization of the Compounds

The matches of the retention time and mass spectra against the calibration standard were employed in the identification and quantization of the PAHs compounds. Internal standardization method was adopted for the quantitation used in this study. This was done through the employment of acenaphthene-d_{10} (internal standard for fluorine, acenaphthene, acenaphthylene and naphthalene), while phenanthrene was adopted as the internal standard for anthracene, fluoranthene, phenanthrene and pyrene. Chrysene-d_{12} was employed for chrysene and Benz(a)anthracene. Perylene-$_{d12}$ was used for Benzo(g, h, i)perylene, Benzo(k)fluoranthene, Benzo(a)pyrene, Benzo(b) fluoranthene, Indeno(1,2,3-cd)pyrene and Dibenz(a, h)anthracene. The ratio of the peak height (substance) subjected to quantification and that of the matching standard (internal) was adopted in calculating the quantification.

5.3 RESULTS

Results of the various hydrocarbon components in Bonny light crude oil utilized that was determined by gas chromatography are shown in Table 5.1. The mean values of residual polycyclic aromatic hydrocarbon content in bottles stimulated and augmented with different concentrations of sewage sludge and bacteria are presented in Table 5.2. Results presented in Table 5.3 show varying and significant reductions in TPH owing to the effect of sewage sludge employed as biostimulant.

The chromatograms showing the PAHs contents of the Bonny light crude oil before and after bacterial degradation are presented in Figures 5.3–5.5. PAHs reductions due to efficient enhancement of the bacterial inoculant can be evidently seen (Figures 5.4 and 5.5).

The results generated on the PAHs components of 5% (1,065.9733 mg/kg), 10% (1,726.5850 mg/kg) and 15% (1,865.6340 mg/kg) SEO contaminated levels before bacterial degradation are presented in Table 5.4. At the fifth week, *Pseudomonas aeruginosa* and *Alcaligenes feacalis* coculture significantly influenced PAHs degradation on all the SEO contamination levels ($P<0.05$). Specifically, this bacterial inoculant enhanced PAHs reductions (170, 384.4 and 371.7 mg/kg) on 5%, 10% and 15% SEO contamination levels, respectively. These significant PAHs reductions are presented in Table 5.5. Further PAHs reductions (64.9, 29.7 and 103.5 mg/kg) were recorded on 5%, 10% and 15% SEO contamination levels at the tenth week when the experiment was terminated (Table 5.5).

TABLE 5.1
Polycyclic Aromatic Hydrocarbons Contents of Bonny Light Crude before Degradation

Hydrocarbon Component	Concentration (mg/L)
Naphthalene	8.77
Benzo (a) pyrene	389
2-Methyl naphthalene	46.4
Benzo (a) anthracene	342
Acenaphthylene	67.4
Acenaphthene	91.1
Chrysene	302
Dibenzo (a, h) anthracene	419
Fluorine	280
Phenanthrene	310
Benzo (b) fluoranthene	416
Anthracene	354
Benzo (g, h, I) Perylene	392
Pyrene	410
Indeno (1, 2, 3-d) Pyrene	330
Fluoranthene	354
Benzo (k) fluoranthene	322
Total	4,833.67

Chromatogram results generated from the TPH concentrations present in 5% (9,934 mg/kg), 10% (10,016 mg/kg) and 15% (10,379 mg/kg) SEO contamination levels before bacterial degradation are depicted in Figures 5.6–5.8, respectively. Results indicate that at the fifth week, *Pseudomonas aeruginosa* and *Alcaligenes faecalis* coculture influenced significant TPH degradation ($P<0.05$). Specifically, this bacterial inoculant bioaugmented with the varying SEO contaminated soil (5%, 10% and 15%) enhanced TPH reductions (756, 1,577 and 1,584 mg/kg) respectively. Chromatograms showing the significant reductions attained at the fifth week are shown in Figures 5.9–5.11. However, at the tenth week when the experiment was terminated, the bacterial inoculant had significantly enhanced further TPH reductions (305, 250 and 516 mg/kg) on 5%, 10% and 15% SEO contamination levels respectively (Figures 5.12–5.14).

5.4 DISCUSSION

The need for enhancing indigenous and genetically engineered hydrocarbonoclastic microorganisms with organic amendments that can adequately provide limiting nutrients which in turn stimulate their optimal performance for any bioremediation project has been extensively established in the literature. According to Awodun (2008), the employment of organic amendments with a view to achieving bioremediation and reclamation of hydrocarbon impacted environments is in agreement with

TABLE 5.2
Mean Values of Residual Polycyclic Aromatic Hydrocarbon Content (mg/L) in Bottles Stimulated and Augmented with Different Concentrations of Sewage Sludge and Bacteria

Hydrocarbon Components (mg/kg)	Varying Concentrations of Sewage Sludge (g/L)		
	3.22	6.44	12.88
Acenaphthylene	51.22	65.05	75.90
Benzo (a) pyrene	16.85	14.65	20.70
2-Methyl naphthalene	08.31	09.88	13.44
Dibenzo (a, h) anthracene	02.10	08.91	11.56
Acenaphthene	03.38	63.85	25.47
Anthracene	40.05	53.45	68.35
Benzo (g, h, I) Perylene	02.89	14.25	09.44
Fluorine	21.40	58.15	56.80
Phenanthrene	25.15	90.60	73.05
Fluoranthene	50.03	51.20	64.45
Indeno (1, 2, 3-d) pyrene	02.98	06.18	09.37
Benzo (k) fluoranthene	20.45	30.90	33.25
Pyrene	21.78	22.30	28.30
Benzo (a) anthracene	31.75	23.60	21.80
Chrysene	21.74	73.65	68.60
Benzo (b) fluoranthene	23.50	19.15	36.75
Naphthalene	07.85	03.85	04.71

TABLE 5.3
Mean Values of Residual Total Petroleum Hydrocarbons (mg/L) Obtained before and after Bacterial Degradation of Bonny Light Crude Using Different Concentrations of Sewage Sludge as Biostimulant

Varying Concentration of Sewage Sludge (g/L)	TPH (mg/L) Degradation (Before)	TPH (mg/L) Degradation (After)
3.22	48,597	4,572.0
6.44	48,597	6,760.0
12.88	48,597	7,243.5

international convention on waste utilization. The choice of employing hydrocarbonoclastic bacteria for possible remediation of the hydrocarbon contents in this study is premised on the submissions of Srivastava et al. (2015), regarding their ability to withstand unfavorable environmental conditions and grow under controlled conditions. Again, it has been documented by Mosa et al. (2016) that bacteria have got excellent biosorption owing to the teichoic acid obtainable on their cell wall,

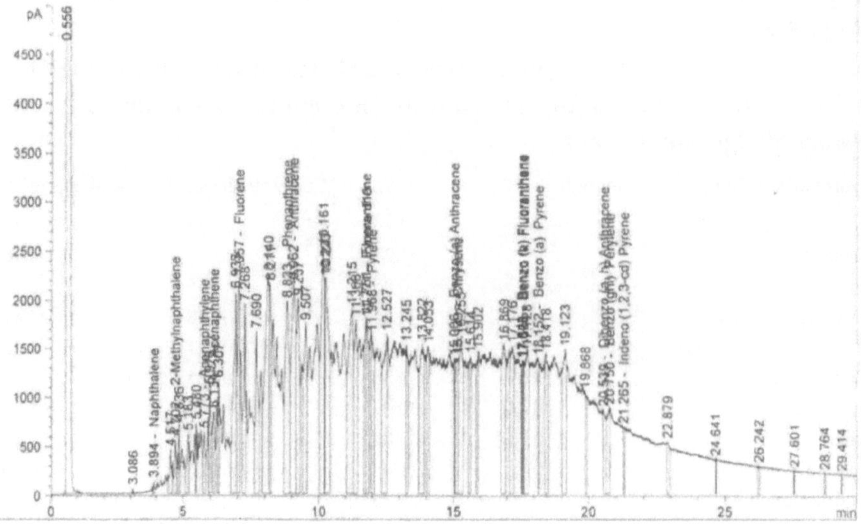

FIGURE 5.3 Chromatogram depicting different fractions of un-degraded Bony light crude oil.

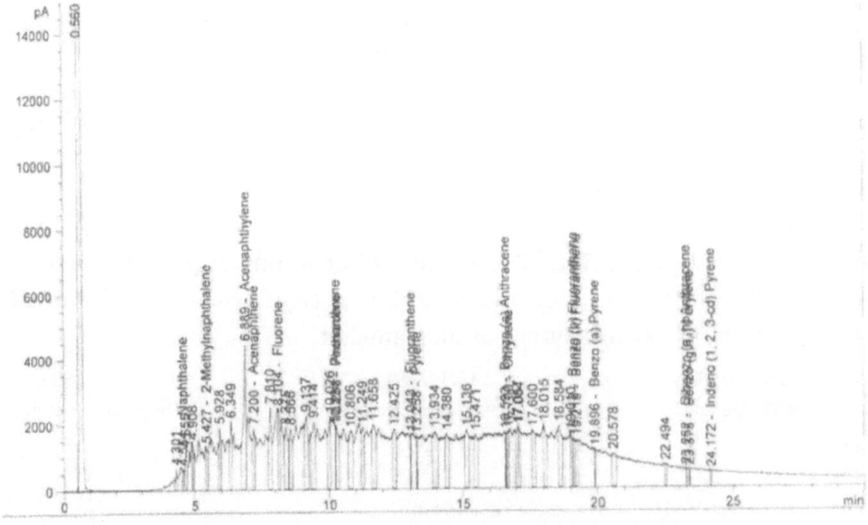

FIGURE 5.4 Chromatogram displaying different fractions of un-degraded Bony light crude oil enhanced with 3.22% sewage sludge and bacteria.

tend to be steadier and survive better when they are found in mixed cultures for any bioremediation project.

PAHs have been extensively documented in the literatures as ubiquitous environmental contaminants. When found or released into environmental media and upon injection into the food chain, many authors (Abu and Ogiji, 1996; Adieze et al., 2003;

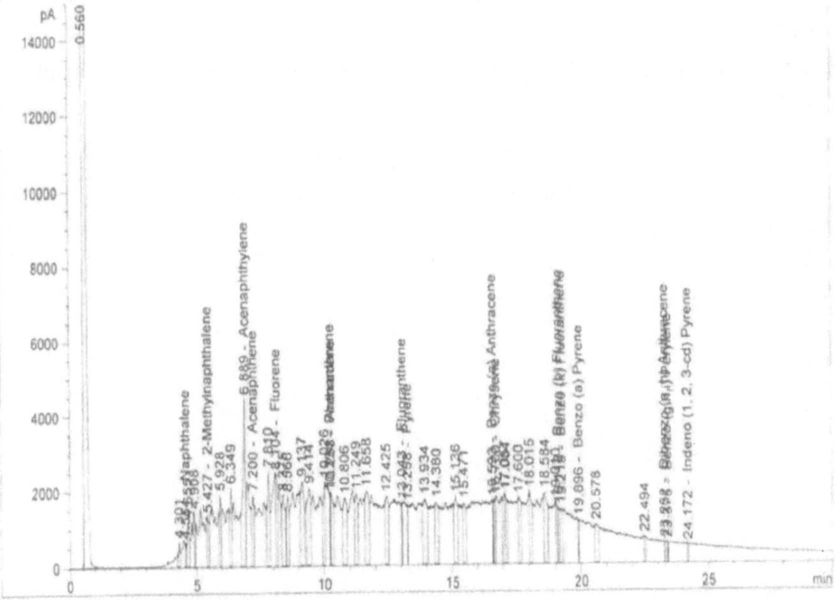

FIGURE 5.5 Chromatogram depicting different fractions of un-degraded Bony light crude oil enhanced with 12.8% sewage sludge and bacteria.

TABLE 5.4
Polycyclic Aromatic Hydrocarbons Contents of SEO Contamination Levels before Degradation

Hydrocarbon Components	SEO Contamination Levels (mg/kg)		
	5%	10%	15%
Dibenzo [a, h] anthracene	17.8087	18.6788	12.9690
Benzo [k]fluoranthene	71.6423	14.2141	174.1245
Acenaphthylene	165.6564	146.4636	286.7352
Fluoranthene	39.1378	154.8751	27.3495
Acenaphthene	119.8004	159.9880	306.9489
Indeno [1, 2, 3,c. d] pyrene	18.7666	12.1187	10.8305
Fluorene	104.3942	192.6765	160.7632
Benzo [a] anthracene	28.2368	150.1325	67.4873
Anthracene	140.0905	334.1136	212.1509
Naphthalene	139.0226	0.9027	212.3526
Phenanthrene	91.2870	210.1942	146.0150
Pyrene	33.3092	118.7443	11.9124
Benzo [b] fluoranthene	41.8028	119.3219	103.1410
Chrysene	14.035	38.5171	41.3368
Benzo [a] pyrene	40.9865	55.4440	91.4574
Total	1065.9733	1726.5850	1865.6340

TABLE 5.5
Polycyclic Aromatic Hydrocarbons Reductions Achieved after Bacterial Degradation

	PAHs Reductions (mg/kg)		
	5% SEO	10% SEO	15% SEO
Fifth week	170	384.4	371.7
Tenth week	64.9	29.7	103.5

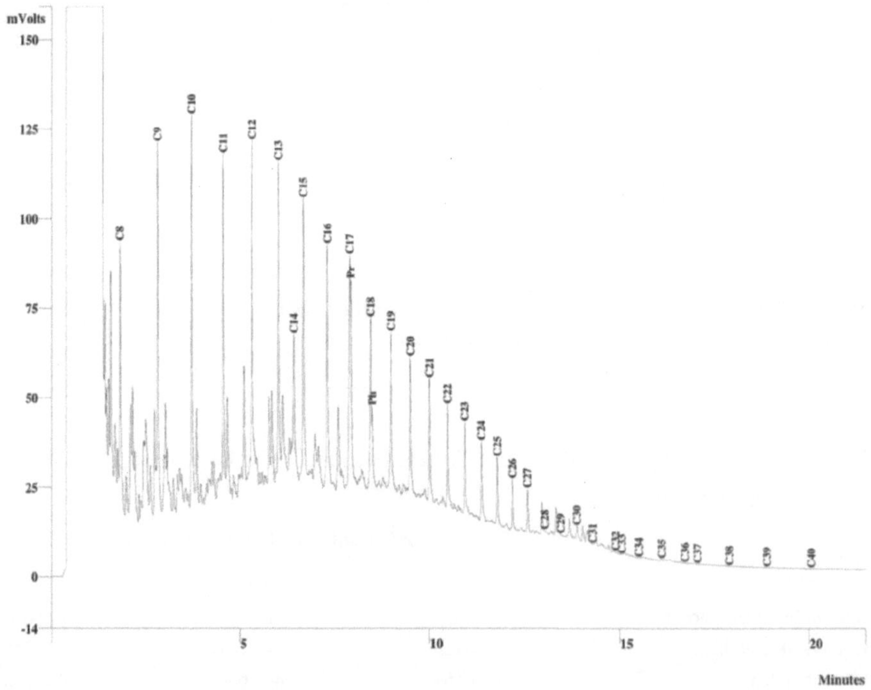

FIGURE 5.6 Chromatogram showing the TPH fractions of 5% SEO contamination level before bacterial degradation.

Hall and Oris, 1991; Potter, 1994) have reported its devastating health implications. Numerous hydrocarbons that are classified as being carcinogenic are ultimately toxic to humans (Thomas, 1990; Vandermeulen et al., 1985; Alonge, 1988; King et al., 2002; Bartha, 1986). Amund and Ogiri (1990) have equally documented the toxic effects of oil on livestock and crops.

In this study, PAHs that are of serious health hazards were significantly reduced through the action of bacterial formulation and biostimulants used. All the PAHs documented by Adieze et al. (2003) ranging from Benzo (g, h, i) perylene, Benzo (a) pyrene, Benzo (b) fluoranthene, Chrysene, Dibenz (a, h) anthracene, Indeno (1,2,3 –d) pyrene and Benzo (a) anthracene are detected in this study. Interestingly, all

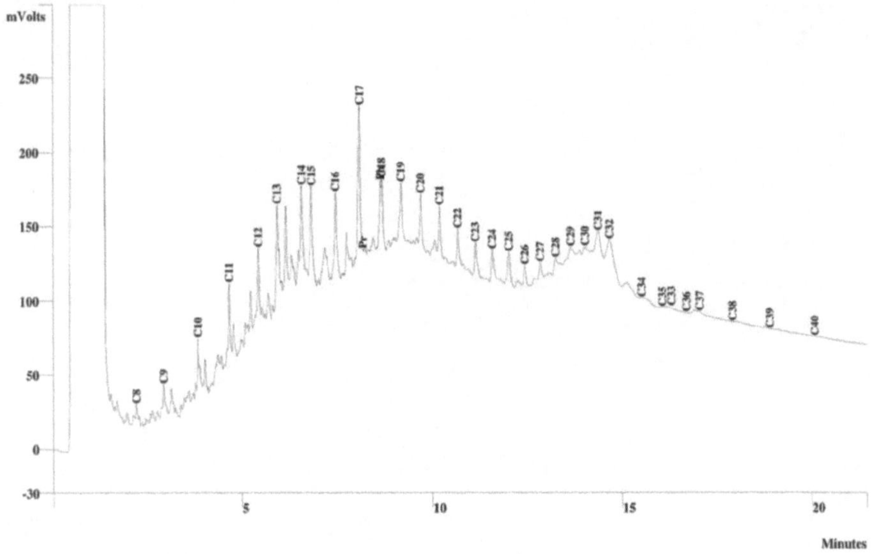

FIGURE 5.7 Chromatogram showing the TPH contents of 10% SEO contamination level before bacterial degradation.

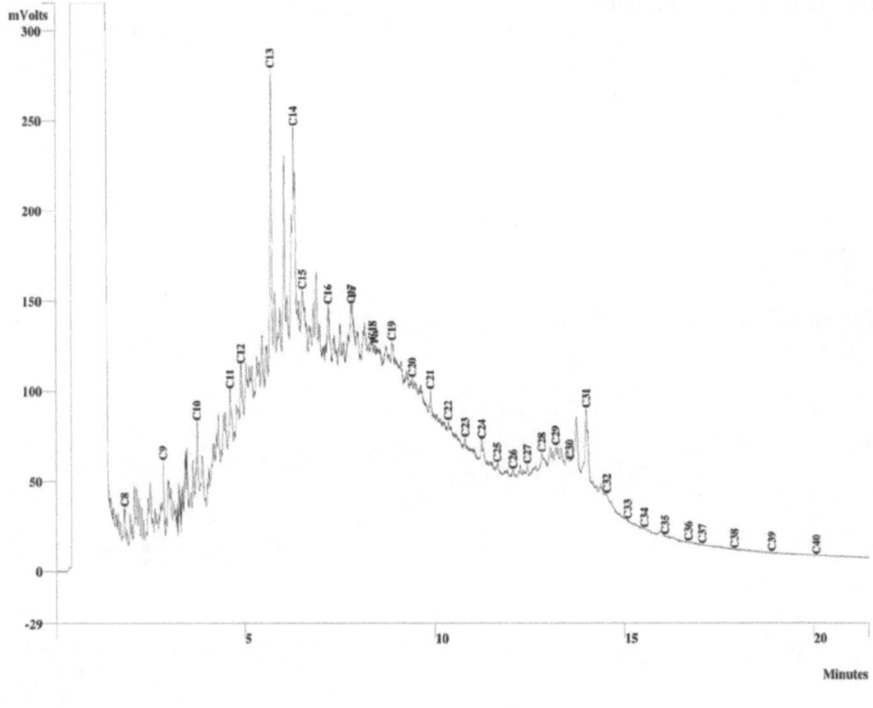

FIGURE 5.8 Chromatogram showing the TPH contents of 15% SEO contamination level before bacterial degradation.

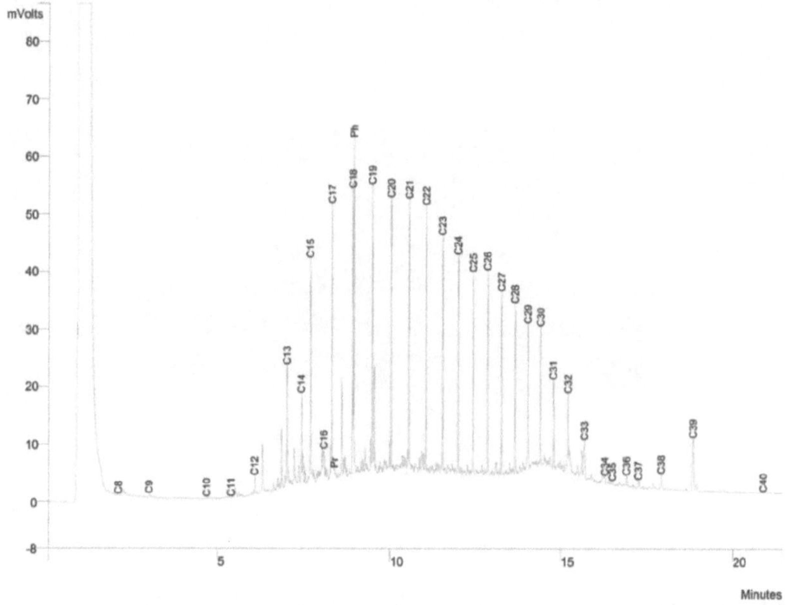

FIGURE 5.9 Chromatogram showing the degraded TPH fractions of 5% SEO contamination level amended with compost at the fifth week using *Pseudomonas aeruginosa* and *Alcaligenes faecalis* coculture.

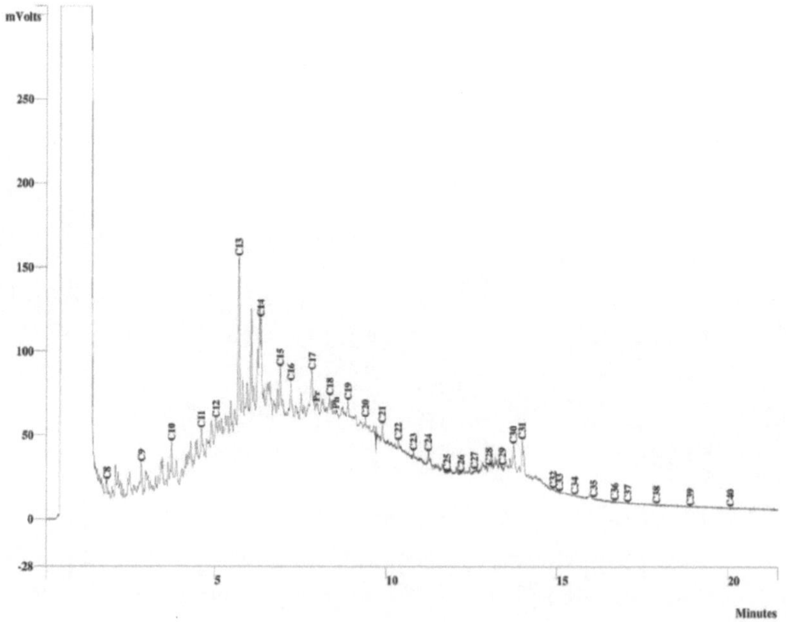

FIGURE 5.10 Chromatogram showing the degraded TPH fractions of 10% SEO contamination level amended with compost at the fifth week using *Pseudomonas aeruginosa* and *Alcaligenes faecalis* coculture.

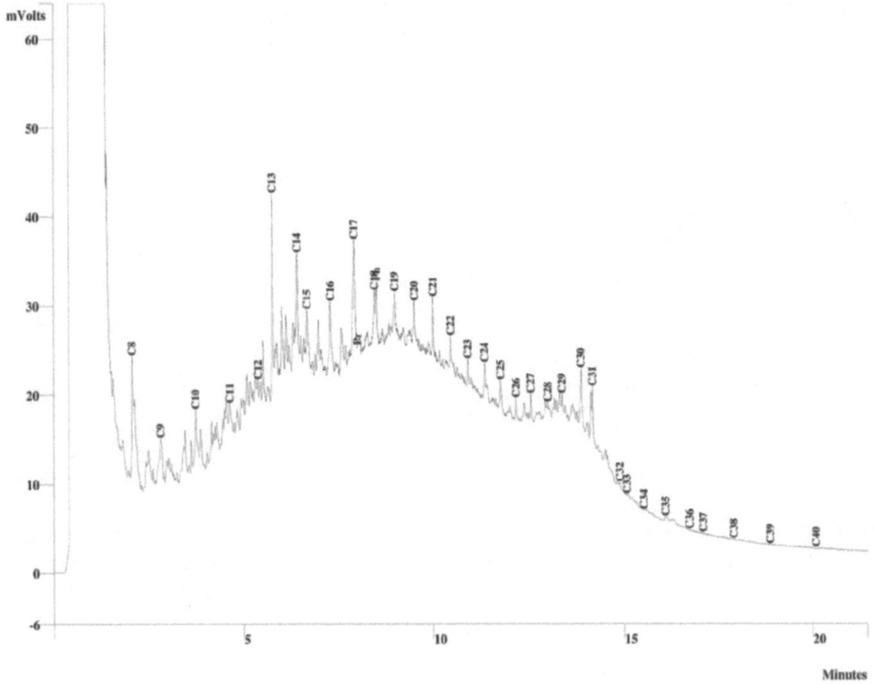

FIGURE 5.11 Chromatogram showing the degraded TPH fractions of 15% SEO contamination level amended with powdered CPH only at the fifth week using *Pseudomonas aeruginosa* and *Alcaligenes faecalis* coculture.

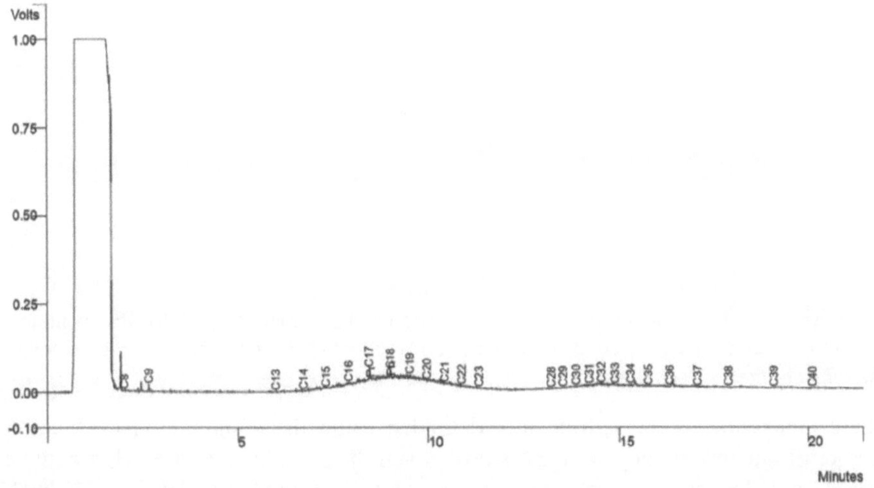

FIGURE 5.12 Chromatogram showing the degraded TPH fractions of 5% SEO contamination level amended with compost at the tenth week using *Pseudomonas aeruginosa* and *Alcaligenes faecalis* coculture.

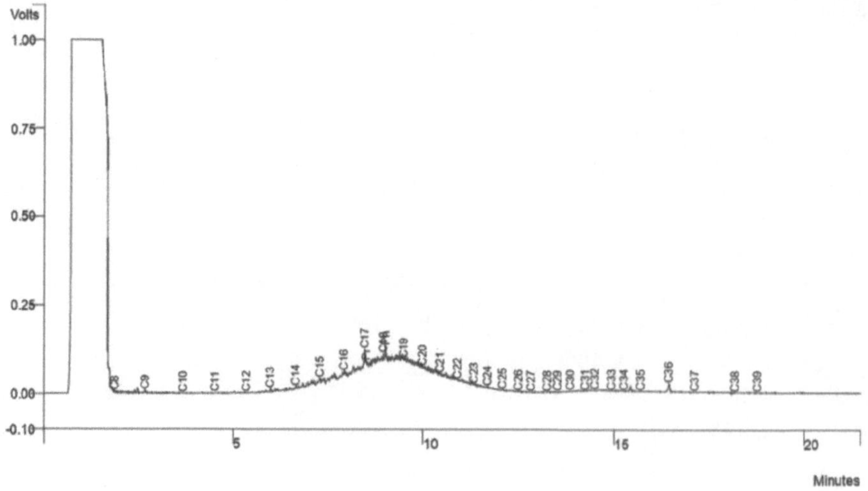

FIGURE 5.13 Chromatogram showing the degraded TPH fractions of 10% SEO contamination level amended with compost at the tenth week using *Pseudomonas aeruginosa* and *Alcaligenes faecalis* coculture.

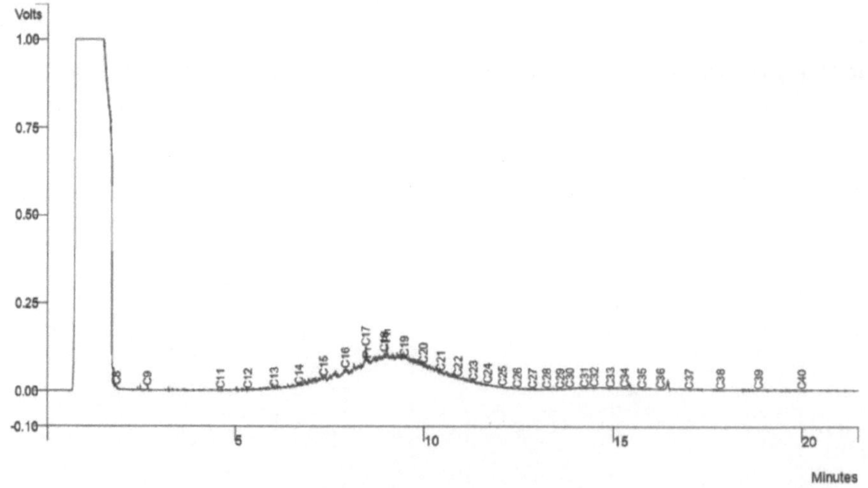

FIGURE 5.14 Chromatogram showing the degraded TPH fractions of 15% SEO contamination level amended with compost at the tenth week using *Pseudomonas aeruginosa* and *Alcaligenes faecalis* coculture.

these compounds were significantly degraded using the combination of bacteria and biostimulants employed in this study. A significant reduction of PAHs fractions was recorded when the lowest concentration (3.22%) of the biostimulant was added. However, with increase in the concentration of the biostimulant employed in this study, anthracene and 1,2,3-d pyrene which are critical PAHs that exert serious hazards were significantly reduced.

The utilization and subsequent reduction of PAHs by the coculture of *Pseudomonas aeruginosa* and *Alcaligenes faecalis* in this study is in agreement with the submission of Ogunbayo et al. (2012) that these constituents of petroleum hydrocarbons are known as substrates that support the growth of hydrocarbonoclastic microorganisms. The bacterial inoculant employed significantly enhanced the degradation of PAHs on 15% SEO contaminated soil from the initial concentration (1,865.6340 mg/kg) to 103.5 mg/kg when the biodegradation assay conducted in this study was terminated. The significant PAHs degradation obtained in this study are in line with the submissions of Lakshmi and Velan (2011) and Singha et al. (2017), on the capability of *Alcaligenes faecalis* to actively mineralize the PAHs contents of petroleum hydrocarbons leading to its degradation.

The ability of *Pseudomonas aeruginosa* and *Alcaligenes faecalis* to produce biosurfactant as reported by Wyrwas et al. (2011) must have enhanced effective bioavailability of PAHs which led to the significant reductions attained in this study. The catabolic pathways in *Pseudomonas* species which enable the degradation of numerous aromatic pollutants most especially naphthalene, without having to synthesize a large number of different enzymes have been reported by Houghton and Shanley (1994). Again, Ismail et al. (2014) have implicated species of *Pseudomonas* as one of the most prolific hydrocarbonoclastic bacteria that have been widely found in hydrocarbon impacted environments in Nigeria.

The significant reductions of TPH content of all the SEO contamination levels occasioned by the coculture of *Pseudomonas aeruginosa* and *Alcaligenes faecalis* in this study can be attributed to their being enhanced optimally by the nutrients obtainable in the biostimulants employed. This result is in agreement with the reports of several authors (Agbor et al., 2012; Agyarko and Asiedu, 2012; Romanus et al., 2015) on the ability of organic amendments to release significant nutrients needed for optimum microbial metabolism into the soil when applied thereby bringing about significant TPH degradation.

As reported by Igwo-Ezikpe et al. (2009), the ability of *Pseudomonas aeruginosa* and *Alcaligenes faecalis* to produce biosurfactants when enhanced optimally with suitable nutrients might have led to the significant reductions of TPH achieved in this study. Diaz (2008) did report that enhancement of biodegradation process gives rise to the production of biosurfactant which might have ultimately helped the TPH degradation rate recorded in this study. Many authors (Vahaboglon, et al., 1996; Jorgensen et al., 2000; Akpe et al., 2013) have linked the ability of gram negative bacteria to utilize hydrocarbons with the presence of porins on their cell wall and availability of plasmid-borne or chromosomal gene meant for degradation of compounds. The duo of *Pseudomonas* and *Alcaligenes* are among the commonest genera of bacteria that are prolific hydrocarbon degraders (Balba et al., 1998). The SEO degradative prowess demonstrated by *Pseudomonas aeruginosa* and *Alcaligenes faecalis* coculture in this study has further substantiated the submissions of these authors.

5.5 CONCLUSION

Due to the results obtained in this study, the adoption of the practical utilization of sewage sludge and other wasted nutrient-rich materials to achieve bioremediation is advocated for the reclamation of related hydrocarbon polluted agricultural farmlands. This is particularly so because the nutrients that are present in sewage sludge and other organic materials employed are loosely bound and can be readily available for microbial utilization in bioremediation and effective uptake by crops involved in phytoremediation. It is worthy of note that tribal differences exist even in the world of microorganisms. It is much better and more economical to further screen indigenous bacteria with the right chemical appetite for hydrocarbons for use in mitigating pollution. This is a food for thought for government and multinational oil companies operating in Nigeria.

ACKNOWLEDGMENTS

The success of this research is attributed to the wonderful support and analytical assistance offered by Shell Petroleum Development Company, Port Harcourt, Nigeria. Encouragement from the pioneer Vice Chancellor of Federal University Dutse, Nigeria in person of Prof. J.D. Amin is also highly acknowledged.

REFERENCES

Abioye, O. P., Agamuthu, P. and Abdul-Aziz, R. A. (2012). Biodegradation of used motor oil using organic waste amendment. Hindawi Publishing Corporation. *Biotechnology Research International*, 2012, Article ID 587041, 8 p Doi: 10.1155/2012/587041.

Abu, G. O. and Ogiji, P. A. (1996). Initial test of a bioremediation scheme for the clean up of an oil polluted water body in a rural community in Nigeria. *Bioresources Technology*, 58: 7–12.

Adegunloye, D.V and Olorunnusi, T. M. (2016). Production of compost from cocoa pod wastes and animal dung. *Nigerian Journal of Microbiology*, 30 (1): 3184–3191.

Adeleye, A. O. and Yerima, M. B. (2019). Isolation and identification of spent engine oil utilizing bacteria from mechanic village Dutse, Jigawa State. *Paper Presented at the 42nd Conference of the Nigerian Society for Microbiology*, Crawford University, Igbesa, Ogun State.

Adieze, I. E., Nwabueze, R. N. and Onyeza, G. D. C. (2003). Effect of poultry manure on the microbial utilization of hydrocarbons in oil polluted soil. *Nigerian Journal of Microbiology*, 17(1): 12–16.

Agbor, R. B., Ekpo, I. A., Kalu, S. E., Bassey, I. U., Okoi, E. P and Ude, E. O. (2015). Growth pattern of two crop species on bio-remediated hydrocarbon polluted soils. *Academic Journals*, 10 (2): 58–63.

Agbor, R. B., Ekpo, I. A., Udofia, U. U., Okpako, E. C. and Ekanem, E. B. (2012). Potentials of cocoa pod husk and plantain peels in the degradation of total petroleum hydrocarbon content of crude oil polluted soil. *Archives of Applied Science Research*, 4 (3): 1372–1375.

Agyarko, K. and Asiedu, E. K. (2012). Cocoa pod husk and poultry manure on soil nutrients and cucumber growth. *Advances in Environmental Biology*, 6 (11): 2870–2874.

Akpe, A. R., Ekundayo, A. O. and Esumeh, F. I. (2013). Degradation of crude oil by bacteria: a role for plasmid-borne genes. *Global Journal of Scientific Frontier Research Biological Science*, 13 (6-1): 21–26.

Alonge, D. O. (1988). Carcinogenic polycyclic aromatic hydrocarbons (PAHs) determined in Nigerian *kundi* (smoked dried meat). *Journal of Science, Food and Agriculture*, 43: 167–173.

Amund, O. O. and Igiri, C. O. (1990). Biodegradation of petroleum hydrocarbons: an environmental perspective. *Microbiology Reviews*, 45 (1): 180–189.

Anonymous. (1994). *Oil Spills and Oil Spill Response*. United States Environmental Protection Agency (USEPA).

Atuanya, E. I. (1987). Effect of wasted engine oil pollution on physical and chemical properties of soil. a case study of waste oil contaminating delta soil in bendel state. *Nigerian Journal of Applied Science*, 5: 155–176.

Awodun, M. A. (2008). Effect of Nitrogen release from rumen digesta and cow dung on soil and leaf nutrient content of gboma (*Solanum macrocarpon* L.). *Journal of Applied Biosciences*, 7: 202–206.

Ayotamuno, M. J,, Kogbara, R. B., Ogaji, S. O. T. and Pobert, S. D. (2006). Bioremediation of a crude oil polluted agricultural soil in Port Harcourt Nigeria. Applied Energy, 83: 1249–1257.

Balba, M. T., Al-Awadhi, N. and Al-Daher, R., (1998). Bioremediation of oil-contaminated soil: microbiological methods for feasibility assessment and field evaluation. *Journal of Microbiological Methods*, 32: 155–164.

Bartha, R. (1986). Biotechnology of petroleum pollutants biodegradation, *Microbial Ecology*, 12: 155–172.

Chorom, M., Sharifi, H. S. and Motamedi, H. (2010). Bioremediation of a crude oil polluted soil by application of fertilizers. *Iranian Journal of Environmental Health Science and Engineering*, 7 (4): 319–326.

Diaz, E. (2008). *Microbial Biodegradation: Genomics and Molecular Biology*. 1st Ed., Caister Academic Press. Retrieved from http://www.horizonpress.com/biod. Accessed on May 22nd 2017.

Ezekoye, C. C., Amakoromo, E. R. and Ibiene, A. A. (2017). Laboratory based bioremediation of hydrocarbon polluted mangrove swamp soil in the niger delta using poultry wastes. *Microbiology Research Journal International*, 19 (2): 1–14.

Hall, A. T and Oris, J. T. (1991). Anthracene reduces reproductive potential and is maternally transferred during long term exposure in fathead minnows. *Aquatic Toxicology*, 19: 249–264.

Houghton, J. E. and Shanley, M. S. (1994). Chapter 1 catabolic potential of pseudomonads: a regulatory perspective. In *Biological Degradation and Bioremediation of Toxic Chemicals*. G. R. Chaudhry. Dioscorides Press: Portland, Oregon, pp. 11–32.

Ibe, S. N. and Ibe, S. C. (1984). Control and dispersion potential of oil spills by bacterial seeding. *Petroleum Industry and Nigerian environment*, Lagos Nigeria.

Ibekwe, V. I., Ubochi, K. C. and Ezeji, E. U. (2006). Effect of organic nutrient on microbial utilization of hydrocarbons on crude oil contaminated soil. *African Journal of Biotechnology*, 5 (10): 983–986.

Igwo-Ezikpe, M. N., Gbenle, O. G., Ilori, M. O., Okpuzor, J. and Osuntoki, A. A. (2009). Evaluation of *Alcaligenes faecalis* degradation of chrysene and diesel oil with concomitant production of biosurfactant. *Research Journal of Environmental Toxicology*, 3 (4): 159–169.

Ijah, U. J. J. and George, B. G. (2000). Potential of yeast isolates in decontaminating crude oil spilled soil. *Nigerian Journal of Science and Technology*, 1: 189–195.

Ismail, H. Y., Ijah, U. J. J., Riskuwa, M.L. and Allamin, I.I. (2014). Biodegradation of spent engine oil by bacteria isolated from the rhizosphere of legumes grown in contaminated soil. *International Journal of Environment*, 3 (2): 63–75.

Jorgensen, K. S., Puustinen, J. and Suortti, A. M. (2000). Bioremediation of petroleum hydrocarbon contaminated soil by composting in biopiles. *Environmental Pollution*, 107 (2): 245–254.

King, S., Meyer, J. S. and Andrew, A. R J. (2002). Screening methods for polycyclic aromatic hydrocarbons in soil using hollow fibre membrane solvent microextraction. *Journal of Chromatography*, 982: 201–208.

Lakshmi, M. B. and Velan, M. (2011). Biodegradation of the toxic polycyclic aromatic hydrocarbon, phenanthrene by an indigenously isolated *Alcaligenes faecalis* MVMB1 strain. *International Proceedings of Chemical, Biological and Environmental Engineering*, 6: 440–444.

Mosa, K. A., Saadoun, I., Kumar, K., Helmy, M. and Dhankher, O. P. (2016). Potential biotechnological strategies for the cleanup of heavy metals and metalloids. *Frontiers in Plant Science*, 7 (303): 1–14.

Mukred, A. M., Hamid, A. A., Hamzah, A. and Yusoff, W. M. W. (2008). Growth enhancement of effective microorganisms for bioremediation of crude oil contaminated waters. *Pakistan Journal of Biological Sciences*, 11 (13): 1708–1712.

Nkereuwem, M. E., Edem, I. D. and Fagbola, O. (2010). Bioremediation of oil polluted soils with organominerial fertilizer (OMF) and Mexican sunflower (*Tithonia diversifolia*). *Nigerian Journal of Agriculture, Food and Environment*, 6 (1&2): 13–30.

Odokuma, L. O and Dickson, A. A. (2003). Bioremediation of crude oil polluted tropical rain forest soil. *Global Journal of Environmental Science*, 2 (1) 29–40.

Ogunbayo, A. O., Bello, R. A. and Nwagbara, U. (2012). Bioremediation of engine oil contaminated site. *Journal of Emerging Trends in Engineering and Applied Sciences*, 3 (3): 483–489.

Potter, D. (1994). Molecular isometry of DNA adducts in rainbow trout (*Oncorrhydcus mykiss*) exposed to benzo (a) pyrene by different routes. *Archives of Toxicology*, 69: 1–7.

Romanus, A. A., Ekundayo, A. O., Aigere, S. P. and Okwu, G. I. (2015). Bacterial degradation of petroleum hydrocarbons in crude oil polluted soil amended with cassava peels. *American Journal of Research Communication*, 3 (7): 99–118.

Singha, L. P., Kotoky, R. and Pandey, P. (2017). Draft genome sequence of *Alcaligenes faecalis* BDB4, a polyaromatic hydrocarbon-degrading bacterium isolated from crude oil-contaminated soil. *Genome Announcements*, 5 (48): e01346–e01317.

Srivastava, S., Agrawal, S. B. and Mondal, M. K. (2015). A review on progress of heavy metal removal using adsorbents of microbial and plant origin. *Environmental Science and Pollution Research*, 22 (20): 15386–15415.

Swannell, R., Lee, K. and MacDonald, M. (1996). Field evaluation of marine oil spill bioremediation. *Microbiology Reviews*, 60 (2): 342–365.

Thomas, P. (1990). Teleost model for studying the effects of chemicals on female reproductive endocrine function. *Journal of Experimental Zoology: Supplement*, 4: 12–138.

United State Environmental Protection Agency (USEPA). (2003). *Method 8015C (SW-846): Nonhalogenated Organics Using GC/FID. Revision 4.* Washington, DC. Retrieved from https://www.epa.gov/esam/epa-method-8015d-sw-846-nonhalogenated-organics-using-gcfid.

Vahaboglon, H., Dodanli, S. and Oztrurk, R. (1996). Characterization of multiple antibiotic resistant *Salmonella typhimurium* strains, molecular epidemiology of PCR-1-producing isolates and evidence for nosocomial plasmid exchange by a clone. *Journal of Clinical Microbiology*, 34: 2942–2946.

Vandermeulen, J. H., Foda, A. and Stuttard, C. (1985). Toxicity and mutagenicity of some crude oil distillates and their water solution fractions. *Water Resources*, 19:1282–1289.

Wilfred, F. M. R., Milner, M. G., Jones, D. M., Lee, K. and Daniel, F. (2002). Robust hydrocarbon degradation and dynamics of bacterial communities during nutrient enhanced oil degradation. *Applied and Environmental Microbiology*, 68 (11): 5537–5548.

Wyrwas, B., Chrzanowski, L., Lawniczak, L., Szulc, A., Cyplik, P., Bialas, W., Szymanski, A. and Holderna-Odachowska, A. (2011). Utilization of Triton X-100 and polyethylene glycols during surfactant-mediated biodegradation of diesel fuel. *Journal of Hazardous Materials*, 197: 97–103.

Yongabi, K. A., Ijah, U. J. J and Elijah, R. I. (2006). Potential of *Pleurotus ostreatus* and urea as biocatalyst in the biodegradation of oil contaminated soil. *A Paper Presented at the seminar organized by the International Organization for Biotechnology (Senegal).* Washington, DC.

6 Environmental Reclamation in Nigeria
Past, Present, and Future

Beckley Ikhajiagbe, Geoffrey O. Anoliefo, and Isaac J. Adekunle
University of Benin

Ogochukwu F. Olise
University of Benin
Anchor University

Saheed Ibrahim Musa
Admiralty University of Nigeria

CONTENTS

6.1 INTRODUCTION

Environmental reclamation, which involves land surface changes such as soil cleaning processes, enhancements, replacement, and re-vegetation to restore deteriorated habitats to their former beneficial past (Sinton, 2011), has been a necessary wonder throughout history. This chapter examines man's commitment to cleaning up and reclaiming the world, from which he has reaped the greatest benefits, regardless of his feelings about other species (plants, microorganisms, and animals), in the past, present, and future. As this analysis focuses on showing the past and present trends

DOI: 10.1201/9781003180289-6

as well as the possible frontline problems in Nigeria, there is an urgent need for concordance between human sustenance and environmental preservation as a pre-requisite for nature's recovery.

The earth is home to a considerable number of life species. Incidentally, earth is most greatly influenced by man's activities. Overpopulation, pollution, burning, combustion of fossil fuels, and deforestation are all consequences of human interventions on the physical environment. Changes like these would eventually impact unfavorable climatic changes, soil depletion, and poor air and water quality. Furthermore, as the human population continues to grow toward a total of 10 billion people, these concerns are resurfacing. It has been agreed that man has the ability to properly call an end the situation, and the situation is currently an emergency – an unparalleled global emergency (Stephen, 2013). Given the current state of environmental degradation, both at sea and on land, as well as air pollution, the fate of global habitats is in jeopardy unless concerted efforts to restore depleted earth are made. Environmental reclamation is one of several ways to deal with fixing this situation which has happened in time.

Contaminants are removed from the environment during remediation. Remediation is any process that restores land resources to their previous state, or pristine condition. Usually the process is guided by protocols and laid down rules. However, where none of the required statutory guidelines exist, remedial intervention is usually subject to a variety of regulatory criteria; some of which center basically on risk assessment of such contaminant on human and ecosystem health. By eliminating toxins and pollutants from surface water, groundwater, sediment, soil, and other aspects of the ecosystem, environmental remediation protects the environment and, indirectly or directly, human health.

Human activities on earth significantly impact environmental sustainability. Overexploitation of earth's resources usually places much strain on natural habitats and resources, and the impact on cycles and other ecosystem services can be very enormous (Millennium Ecosystem Assessment, 2005). Land resources have also been severely strained as a result of urban and industrial expansion. A research carried out on the effects of soil remediation via soil exploration of sea shore front systems in Ikate and Ilubirin in Lagos, Nigeria (Sridhar et al., 2019) revealed evidences of well-reduced water concentration than that of WHO (2017) benchmarks. Ikhajiagbe et al. (2020b) reported that *Vernonia amygdalina*, a common medicinal shrub in Nigeria, has the capacity to hyperaccumulate heavy metals in its leaves. Soil remediation of heavy metals by *Chromolaena odorata* has been reported (Anoliefo et al., 2017; Ikhajiagbe, 2016a, b; Ikhajiagbe and Musa, 2020). The restoration of a petroleum hydrocarbon-polluted soil by common lawn ruderals like Bahama grass and Guinea grass has been reported by Ikhajiagbe et al. (2017a). The contaminants accumulated in the plant biomass and soils became devoid of the contaminants after the plants were removed. The pollutants were metabolized or biotransformed into less harmful or environmentally friendly compounds in some instances.

Land recovery rules had already been established, based on benchmarks derived from the original and basic concentrations, as well as statutory limits. Contaminants pose a threat to people and the world in general, considering the myriad of human activity and the concomitant impact on his immediate environment. However, estimating the pollutant's potential environmental risk is normally difficult due to the matrix's complexity, a lack of reasonable data on the pollutant, the pollutant's fate, and its concomitant effect on the ecosystem (Cachada et al., 2016).

Mankind developed around 200,000 years ago as a species. By 2050, our children or grandchildren will be living on a planet with an additional population of around 9 billion (Stephen, 2013). Man's present condition has arisen through various evolutionary and social-shaping "events," most strikingly the cultivating revolt, the logical change, the cutting-edge change, and the general quest for human well-being. By the basic signs of the results of development and civilization beginning to surface, not the least of these was on water, the demand for water – the water drunk by man and that needed for food production and all of man's consumables – was getting to the peak (Stephen, 2013). It is getting obvious to the global academia system that collection of carbon dioxide (CO_2), methane (CH_4), and ozone hurting substances of the environment – due to growing agricultural businesses, utilization of land and manufacturing activities, handling and transportation – adversely affects the atmosphere (Stephen, 2013); in this manner, man is faced with significant issues at hand, of which the year 1998 had been recorded as the hottest year so far. Ten most harsh periods in terms of temperature increase on record have occurred since 1998 (Stephen, 2013). Man's activities had started to change these parts. Release of CO_2 alters the climate. The overbearing use of water had begun to change the world's water environment. Increased barometrical and sea surface temperature had started to modify the cryosphere; man's growing use of land for farming, urban networks, boulevards, and mining had started to adjust the biosphere, or to put it another way, man had begun to alter the atmosphere (Stephen, 2013).

Nigeria, for example, has a population of over 200 million, and as this continues to grow, so do more water, sustenance, land, vehicle movements, and undeniably higher energy requirements. Man is thus reviving the rate at which man is changing the atmosphere. Stephen (2013) opined that the enthusiasm for soil sustenance is approaching two-fold – by any rate – in the year 2050, and this will triple – at any rate – before the end of the present century. This scenario raises concern from several people about the Nigeria's remaining tropical rainforests. Approximately 70% of the population will live in urban structured communities by 2050. This present age will experience rapid development in urban locations, in the same way as the rise of recently non-existent urban areas. The primary arrangement left to us is to radically and all-inclusively change the behavior at each level of environmental recovery. To put it plainly, man needs to consume less seriously. In addition, radical government action would be needed to bring about such a scandalous change directly. According to the United Nations, the number of inhabitants in Nigeria may likely snowball by the end of 2300. Man needs to urgently undertake drastic action to arrest the ecological catastrophe. In this way, appropriate careful steps should be taken to stop ecological debasement and therefore call for environmental remediation.

6.2 AN OVERLOOK ON THE NIGERIAN ENVIRONMENT

The environment is the totality of the condition within which life forms exist, which is a consequence of communication between dead and life parameters (Barrow, 1993). As a result, our environment may be the totality of all external features that affect man or other organisms throughout life (Ahove, 2001). Mother nature depicts circumstances in which there is no human change in that environment, although the adapted environment is that domain in which recovery needs to be described as one in which critical human modification has taken place. Environment could mean anything to anyone on different occasions, but a more widespread conception is that the environment is all that is outside of a living being. It can currently be all we can see or only experience, implying that the world can be viewed as physical, mechanical, or even normal. It's something that has a real, artificial, or organic impact on life. The environment is a no-do-without for all life forms, especially humans, because it contains everything that man needs for growth. In either case, regardless of the genetic make-up of the organism, the environment may ultimately change the formative structure of a living being. This shows how fundamental the environment can be to life when all is said and done.

The ecosystem encompasses humans, water, air, soil, plants, and other animals, as well as their interactions. The environment comprises of biological components in interaction with themselves and non-biological components and conditions that determine a person's reality, progress, and persistence at any given time. The immediate surrounding of an individual is usually regarded as their environment. One of the major biological factors that affect the sustainability of any environment is the population factor. To what extent do the dwellers occupy an environment without necessarily putting a significant strain on resources? This is also now a problem for many countries including Nigeria.

Nigeria has an estimated population of over 200 million people, and a land area that covers 923,786 km^2, with over 7% of the country's total land area covered by forests (FAO, 2015); while other woodlands cover 2.9% of the total land area, with a total rising stock of 936 million cubic meters. Given the fact that Nigeria's population is rapidly growing, the pressure on the environment for resource provision is very likely to reach a breaking point. When compared with the present condition of the country, the environmental trend in about 60 years prior to this time was nothing compared with what man has today. The economy of the country was very much upheld by agriculture and somewhat less environmental harming exercises. In any case, as the monetary spine floated in favor of oil, there started an unguided increment in urbanization and industrialization which eventually accounted for an expansion in populace. This development likewise brought about numerous individuals halting the agricultural sector because of urbanization and industrialization. The prevalence of oil exploration in the Niger Delta, as well as the crisscrossing oil transportation pipeline from the south-south Geopolitical Zone to all parts of the world, raises concerns, especially in the event of an oil spill.

According to the 2018 Land Degradation Neutrality Target Setting Programme (LDN TSP) study, land degradation is a serious problem in Nigeria that affects the entire country, with consequences such as drought, sand dune, and desertification, which are particularly prevalent in the north, and soil erosion in the south. According to default data given by the UNCCD Secretariat in the form of the LDN TSP, more than 463,360 hectares of forest land was lost between 2000 and 2010.

Despite the vast land mass available in Northern Nigeria, Badejo (1998) claimed that land use change has had an effect on the country's northern ecosystems over time; this has gradually given rise to the fast encroaching dessertification down south, thus occupying areas hitherto occupied by rainforests. An example is in Edo North that is gradually losing its significant forest cover to the fast encroaching-derived savannah ecological zone.

The availability of consistent rainfall is a much more serious issue. Given the fact that soil conditions are highly influenced by rainfall, especially where improved soil moisture enhances humus formation, it is only imperative that the drier and hotter areas are most likely to be low in humus. Secondly, since Nigeria's north basically experiences high levels of solar radiation, soil water loss due to evapo-transpiration is most likely to be higher in this region (Hekstra, 1985). This thus predisposes the soils to water deficit conditions, which hamper plant growth and development. Many plant species (generalists with a wide ecological niche) have been found to survive and flourish in Northern Nigeria in spite of the aforementioned difficulties. Perhaps, apart from genetic capacities for innate tolerance, improved soil conditions which may include sufficient iron concentrations in soils in the north or other improved soil management practices may just turn up to be one of several reasons why select plants would survive therein. However, most southern soils have higher levels of "iron" than is needed, thereby posing the threat of iron toxicity. Crop production is hampered in Benin City, where most of the soils are ferruginous (Musa and Ikhajiagbe, 2020), except for well-adapted crops. Even so, these crops' growth and yield outputs were hampered (Ifie et al., 2020; Ikhajiagbe et al., 2020a). Some researchers have attempted to use a variety of measures to improve crop production in high ferric soils in a number of cases. Ikhajiagbe and Musa (2020) investigated the application of biosynthesized nanoparticles in the enhancement of yield components of rice (*Oryza sativa* var. Nerica) in a ferruginous ultisol and reported enhanced growth responses.

Land deterioration is noticed in mining locales, as portrayed in the Jos Plateau, where tin mining has occurred for few years. This is also reported in numerous territories where extraction of limestone is carried out. Specifically, in Jos Plateau, great expanse of land surface deterioration occurs, bringing about different sizes of various quarries. The farmlands are also possible pollution sites especially with consistent fertilizer overuse, notwithstanding the open indiscriminate disposal metal-based or oil-based waste.

In Lagos alone, there are in excess of eight significant locales for the extraction of sand used for building purposes. Sand digging sites are typical locations in and around towns and in towns everywhere throughout Nigeria. Numerous development companies have started up their own mining sites in Abuja, Minna, Kaduna, Ado-Ekiti, and Okigwe. Notwithstanding, however little the magnitudes and also given that damage is done to soil surfaces during every mining activity, basic problems incidentally happen, when dugged pits at extraction sites collapse, causing a few people to be wiped out (NEST, 1991; also see Table 6.1).

The vegetative cover basically helps with decreasing the effect of land degradation. This is feasible by the obstacle confined by the vegetation, mostly by stem grass-cover and root structures. The vegetative cover grips firmly the topsoil in one piece and diminishes the pace of crumbling. Anyway, cultivating, logging, development activities, overgrazing by domestic animals, intentional devouring of habitat, and various practices that obliterate plant spread make the soil powerless

TABLE 6.1
Environmental Impact of Fossil Fuel Resources (NEST, 1991)

Mining Activity	General Impacts	Identified Effects
Exploration	Landscape disturbance	Aesthetic degradation of the landscape.
Mineral mining	Land deterioration and ecosystem impairment	Land subsidence, seepage system disruption, erosion, excess water draw-down, and water table reduction and pollution are all examples of land surface degradation.
Processing, haulage storage, and consumption	Gas spills, oil slicks, commotion, and contamination of the air, soil, and water	Thermal stacking of waterways, ozone layer destruction, air, soil, and water acidification, climate change, toxicity hazards to plants and animals, extinction of earthly and marine life, loss of crops and domesticated animals, deterioration of atmospheric visibility, vehicular mishaps, damage to buildings and machinery, nervous disorders, respiratory illnesses, and cardiovascular disease are all examples of the consequences of climate change.

against breakdown which affects the Nigerian environment. Exposure to sun may be expressed as the proportion of solar emission reaching a domain for each unit area, and the portion of sun-based energy that gets to the earth's external environment is known as the solar constant.

6.3 POLLUTION IN NIGERIA

The Nigerian pollution problems have expected broad concern (Tomori, 1977). The Royal Commission on Environmental Pollution in the United Kingdom described pollution "as the introduction of substances by man into the environment or energy prone to cause harm to humans, damages to living assets and ecological systems, destruction to structure or enhancement or impairment with rightful utilization of the environment." Pollution is defined as the exudation (into any atmosphere) of harmful products causing harm to man or other creatures in the environment, according to Section 1(3) of the United Kingdom's Environment Protection Act of 1990. When there is the potential for damage, pollution occurs. Natural contamination may widely be categorized into: (1) Ecological contamination where the earth is polluted much of the time with, for instance, seismic tremors, floods, dry season, and tropical storms, and (2) Man-made contamination, which is realized by human activities. The environmental contamination can in like manner be subdivided into air contamination, water contamination, land contamination, sustenance contamination, upheaval contamination, and radio-dynamic contamination.

Numerous environmental problems in Nigeria are linked to the rapid speed of urbanization, the mentality of people living in urban areas around the world, and poor environmental management. It is difficult to peg the real cost of environmental degradation on daily services to the Nigerian economy. Nigeria's various provinces face a variety of environmental problems in varying degrees. Whereas the trouble of wind-induced soil erosions affects most parts of the north, water-based surface soils erosion affects the south. A number of deep gullies crisscross several cities especially the popular "Queen Ede" Moat in Benin City. The impact of wind erosion is enormous in the north, to the extent that arable lands as well as most farms are submerged by skimming sands which spread over juvenile crop yields. Highways, lodges, and open structures might be totally secured by powerful sand rises peaking presently and reach over 12 m high.

In Nigeria, huge dependence is hinged on annihilation of destitution via gigantic industrialization program (Tomori, 1977), which affects negatively the method for environmental reclamation; and hence, the right methods must be taken to accommodate such contradicting destinations. Due to the current trend of development in Nigeria, clearly the continuous social costs of ecological deterioration peak to an unavoidable magnitude of fast pushing degrees of monetary action and the population's development in the presence of progressively evident cut-off areas in the assimilation limits of the biosphere (Ingo, 1972).

Cars, pesticides, fungicides, mechanized fuel, chemicals, discarded items, and objects related to built structures are the wellsprings of our natural endowment. Ecological population reduction and loss due to overuse of toxic pesticides and herbicides, loss of a few types of plants due to monoculture, and expulsion from territories due to deforestation have now become a global environmental concern (Appannagari, 2017), especially in Nigeria. According to Appannagari (2017), the reasons for environmental pollution are populace development, general increased income and monetary development, nature of financial development, nature of present-day innovation, deforestation, agricultural development, modern advancement, and urbanization. Increased population incurs weight on the environment. Ecological problems associated with increased population result in debasement, disintegration, and desertification due to man's maltreatment of naturally occurring materials.

Before now it was argued that there is a relationship between masses size and per capita damage to the ecosystem, which makes cost of keeping up ecological quality at a given level increase excessive as population snowballs (Paul and John, 1971). The negative impact on the environment by an increasing population cannot be sufficiently taken care of by the present financial-related esteeming system. "The condition is encapsulated whereby in any event sources of contamination are associated with a productive monetary intrigue that agitates an immense and diffuse collection of parties" (Allan, 1971). As far back as 1968, it was opined that Lagos, Benin, Sapele, Onitsha, Aba, Port-Harcourt, Ibadan, Abeokuta, Kano, Zaria and Jos are places where new hardcore machinery are utilized, and it has been evaluated that these territories speak about half of production processes in Nigeria (Aboyade, 1968).

The nature of life, whether it is that of humans, animals, or plants, is dependent on the quality of the environment. Humans, as well as plants and animals, can live longer and experience a higher standard of life in a country with a high-quality climate, allowing for sustainable growth. Despite the fact that global industrialization is putting more pressure on environmental degradation, poverty appears to be the worst offender, especially in developing economies. Poverty in Nigeria is at an all-time high, and it is only one of the many economic challenges that the Nigerian people face. Poor environmental quality, which is widespread in Nigeria, is one of the factors that may be contributing to the current high levels of poverty. As attempts are made to improve environmental quality, poverty levels may rise. Although poor environmental quality has a number of negative consequences for a country, the most significant are those for health and education, both of which have an impact on individual well-being. Poor health, which can be caused by a poor-quality climate, has a detrimental impact on an individual's ability to earn a living.

According to the Brookings Institute's World Poverty Clock, poverty is at an all-time high in Nigeria (Adebayo, 2018). Despite the Nigerian government's efforts over the years to combat high and increasing poverty, it is clear that previous efforts have yielded little result. Poverty drives people to cut down trees, which adds to desertification, soil erosion due to intensive and inadequate land management practices, oil bunkering, and pipeline vandalism, to name a few. Poor environmental quality is not exclusive to Nigeria; it is more prevalent in developing countries as a whole. However, if problems such as poverty are to be a thing of the past in Nigeria and other developing countries, the government must continue to prioritize environmental quality. This is especially true given that achieving zero poverty and creating sustainable cities and communities are two of the United Nations Sustainable Development Goals that Nigeria, along with the rest of the world, must aim for by 2030.

6.4 INDUSTRIAL REVOLUTION IN NIGERIA AND ATTENDANT ENVIRONMENTAL IMPACT

The pre-colonial era, the early post-colonial period, and the events of the mid-1980s are all part of Nigeria's industrial formative stage (Ajayi, 2011). The Nigerian economy prior to 1900, that is, prior to the colonial period, displayed a wide range of art craftsmanship in various clans and terrains, where modern-day processing plant operations were uncommon (Onyemelukwe, 1983). During this time, there was no detrimental effect on the environment.

The citizens of the entity now known as Nigeria were engaged in manufacturing activities prior to the invasion. In the pre-colonial period, that is, before 1900, the various clans and kingdoms in Nigeria had significant craft industries. Bronze casting was practiced in ancient Benin. Early foreigners exported a lot of pottery and other types of art, and they even carted them away during invasions, as we saw with the Benin Invasion of 1892. These manufacturing processes had minimal contributions to environmental pollution.

Since the majority of African countries achieved independence during the 1960s, promoting industrialization became a major challenge for the continent. Industrial development was seen as a way for the continent to achieve self-sufficiency and reduce its dependency on developed economies by the respective governments. Africa's philosophy and values were founded on the assumption that industrialization would turn the continent's economies from conventional agrarian to modern and industrialized. Industrialization was seen as a tool for economic development that would help the continent achieve its macroeconomic goals. The arrival of foreigners at the start of international trade sparked the various types of modern industrialization. Slave trade gave rise to "authorised exchange" (Flint, 1960), in crude materials accessible right now in West Africa. The Nigerian current scene altered after World War II (Adegbola, 1983; Onyemelukwe, 1983); there was the longing for many

African products – crafts as well as vast lands for agriculture, from Europe for crude products following the post-war reconstruction requirements. This led to establishment of several foreign industries including the Niger Company, which preceded the amalgamation of the Northern and Southern protectorates of Nigeria today.

Among the 47 industrial companies in the nation during the pre-1947 era, 21 (44.71%) were busy with production exercises, while 26 (55.3%) establishments were related with the finishing parts of assembling (Onyemelukwe, 1983). Small printing shops and pastry shops made up 15 (31.9%) of the 26 businesses founded at the time. A couple of remote (mostly European) administrators, teachers, and shippers provided significant assistance to these businesses. The post-colonial era saw an unprecedented shift in imports and the beginning of a downward trend in the price of ready-to-use crude materials (Ajayi, 2011). Following Nigeria's independence in 1960, the country experienced an industrial revolution, with a slew of medium and high-tech industries springing up in manufacturing, transportation, and, later, the oil and gas industry.

Nigeria's colonization ushered in the era of modern manufacturing and processing. The government embarked on a robust import substitution policy after independence, during the early post-colonial period. However, rather than local production, this import substitution strategy, which was intended to minimize reliance on foreign trade and save foreign exchange, led to the mere assemblage of foreign-produced goods. Just about 150 medium- and large-scale industrial plants existed at the time of independence, the majority of which were built in the late 1950s. However, by 1965, the number of medium- and large-scale firms had increased to 380, owing to the intensification of the import substitution process and the establishment of firms to produce products previously imported domestically, though still dominated by low-tech light industries (Dare, 2007).

Overpopulation is exacerbated by urbanization, especially in downtown areas. Increased population and citywide construction are used to achieve urbanization. It is represented in Nigeria with urban ghettos with severe environmental consequences. The problem is serious, and it exemplifies the failure of progressive policies to keep up with population growth. Because of the steady development of municipal areas, including the participatory powerlessness of public administrations as well as framework structures which are necessary, keeping up with the pace of growth is important; else environmental conditions in urban areas would eventually break down. Nigeria, like every other developing country, faces obstacles in its growth and efforts to improve the quality of life of its people. The critical economic concerns are the need to promote long-term, rapid economic growth to meet the needs of its large population, as well as the necessity of properly integrating its domestic economy into the global economy in the face of increasing globalization. The consequence of the proliferation of urban centers is a progressive deterioration of the environment. Apart from the fact that lands would have to be vacated to make way for residential space as well as factories to meet the needs of an ever-increasing population, the waste generated as a result would be a worst-case scenario.

The rapid development of non-destructive materials, such as plastics, has resulted in a significant waste discharge problem. Inadequate storm drains, refused dumps in waste lines, and the construction of buildings near or on regular waterways have all been identified as contributing factors to the increasing number of water issues in urban core areas. Environmental problems linked to the expanding improvement of urban ghettos, such as overcrowding in unsanitary housing, poor quality or detachment of basic systems and social management, such as water and sanitation workplaces, and the absence of access courses, are all factors (NEST, 1991). Improvements in urbanization and modern development, combined with poor waste management, have resulted in an alarming level of pollution in Nigeria's land zones.

The effect of an ever-increasing human population on food supply is inevitable. To carry out this goal, the majority of farmers use chemical fertilizers in the production of food in order to replenish the soil. This has left little space for traditional agricultural practices such as crop rotation, bush fallowing, and change cultivation, which will help the land recover naturally. Another reason for the extinction of these old practices may be a lack of land for propagation and crop production, as urbanization has compelled people to use vast amounts of land that could have been used for farming practices to erect houses and start businesses. Due to this situation, farmers have no choice but to use chemical manure to increase the limited amount of available land for crop propagation (FAO, 2001). Inorganic fertilizer use, on the other hand, does more harm than good to the soil; soil may become polluted with metal stabilizers from fertilizers, such as copper, or it can become overloaded with nutrients, causing nutrient toxicity in plants.

With the country's continued urbanization and industrialization, there has been growing concern about air and water pollution. Nigerians are constantly warned about the hazards of gaseous emissions from industrial effluents, car exhausts, and even most recently from generator fumes (European Commission, 2006; Oladipo, 2008a, b). Apart from the bad impact they have on the environment, the effects on human health and longevity are enormous.

6.5 MUNICIPAL WASTE GENERATION AND BURDEN IN NIGERIA

Human population is expected to continue to increase, especially in developing economies such as Nigeria, where population control measures are not given the attention they deserve. If this perceived population growth continues to occur, increase in waste production, carbon emissions, and the appearance of various heavy metals will become inevitable; and these are harmful to human comfort and cause global warming and air pollution. The introduction of hazardous substances and materials as a result of human activities and interventions brings to mind the fact that most environmental issues due to pollution or contamination are almost certainly the result of caused or poorly managed wastes.

The concentration of the Nigerian population in a specific geographic area would result in a significant increase in population. This will lead to the production of waste products such as scrap metals, mechanical waste, and household and agrarian waste. This would pollute the soil, water, and air in the area. As a result, if the current rate of growth persists and Nigeria's population continues to increase, the Nigerian environment will suffer greatly. Waste, according to Oyebode (2018), is any material stream structure discarded by the entire population, and the proportion of waste generated in any city is typically an indicator of the degree of human activities, such as human growth, urbanization, and social change, resource misuse, and unregulated inventive advancement (Adejobi and Olorunnimbe, 2012; Amuda et al., 2014). In Nigeria, issues with municipal solid waste have recently become evident as a result of urbanization and a lack of waste evacuation (Oyebode, 2018).

Regrettably, many people in African countries such as Nigeria now regard the need for innovative waste management systems as a minor issue that may have been overlooked in favor of the more serious and pressing issue of achieving rapid economic growth. This mindset stems, to some extent, from the belief that natural depletion of urban solid waste production is an unavoidable cost of development. Solid waste is described as any garbage, unwanted refuse from a waste treatment plant, a water treatment plant, or an air pollution control office, as well as a variety of materials, including solid, liquid, semisolid, and contained gases, produced by industrial companies, advertising, mining, and rural activities. Solid wastes are pollutants that are left on land or dumped into water sources or the atmosphere for a set period of time and may (or may not) be harmful to humans and nature.

Nigeria produces over 32 million tons of solid waste per year, of which only 20%–30% is collected (Bakare, 2020). The careless dumping of Municipal Solid Waste (MSW) has clogged sewers and drainage systems, choking water bodies. The majority of waste is created by households and, in some cases, by local factories, artisans, and traders, and it litters the immediate area. Lagos State, Nigeria's commercial capital, is Africa's second fastest-growing city and the world's seventh fastest-growing city. The city produces more than 10,000 Mtons of urban waste per day, with a per capita waste generation of 0.5 kg/day. Kano and Rivers States, which have a sizable population, face a similar fate in terms of solid waste management and control. In general, Nigeria's population growth has resulted in exponential economic growth, but at the expense of the environment, as more waste is produced as a result of the need for housing, manufacturing industries, and increased trade volume.

6.6 EFFECTIVE RECLAMATION

Development can be beneficial only when it does not make a nation more susceptible to devastating effects on the environment. When a country's environmental foundations are eroded, the economy suffers, the social fabric deteriorates, and the political system becomes unstable. The world, on the other hand, is a dynamic and interactive structure; within this structure, the environment, earth surface, water bodies, physical and chemical components, and living objects are all included. Reclaiming the environment from possible degradative impacts is therefore imperative. Successful reclamation, according to Sinton (2011), begins with practical preparation and communication. Several businesses should hire capable, skilled environmental experts to help them organize their operations. It all comes down to environmental stewardship. The carrying capacity of supporting organic systems is referred to as sustainability.

Environmental reclamation is the foundation of any clean and healthy environment. It is accomplished by the implementation of a cleanup system and the setting of environmental priorities and objectives. In Nigeria, it is imperative to place the maintenance of the characteristics of any given environment as a priority. Protecting the environment therefore guarantees sustainable ecosystem development and better interactions within the same environment. This is a deliberate action. The various techniques by which the fundamental and extended status of the world is restored so that life can continue even as development progresses, to create an equality where the impact of one is verified by the other, and to maintain a strong domain where life can be gratified even at its peak, are referred to as reclamation or recovery (Sridhar et al., 2019).

Many industries (small, medium, and large) were built in the country in the past without adequate consideration of their environmental aspects and impacts. Operations were carried out in cases where laws and regulations did not exist or, if they did, were insufficient and not quite detailed. As is still the case today, waste was discarded carelessly. As a result, polluted areas have emerged. Almost every municipality has a plethora of waste dumps littering its nooks and crannies. Since polluted sites may have unfavorable health effects on the local population, adequate measures must be taken. There are only two ways to effectively remediate polluted land areas or other contaminated media, such as surface or groundwater.

Direct cleanup of contaminated or impacted areas may result in the isolation, immobilization, or removal of the pollution source. Second, assessing the risks of contaminant exposure to people and considering ways to split the pathways between the contamination source and the people will almost certainly ensure that future waste and contaminant production will be held to a minimum. This strategy could result in evacuations, area isolation, or changes in land use and local population habits. These two approaches are mutually beneficial.

Several environmental regulatory agencies, such as the Environmental Protection Agency, define requirements that specify when environmental remediation is required, while different areas may be affected by additional legislation. Oil spills, for example, also necessitate immediate remediation because they pose an immediate danger to the environment and human life. In Nigeria, the Federal Government Protection Agency (FEPA) Decree 58, which established principles for dealing with the environment, was passed in 1988. There had already been many laws and protests of government concern for environmental protection, including the Mineral Act of 1969, 1973, and 1984, the Oil in Navigable Water Decree of 1968, the Associated Gas Injection Act of 1969, and the Chad Basin Development Act of 1973. These laws and acts were enacted to address specific and visible environmental issues. They had a limited reach and were spatially constrained.

Proclamation No. 58 of 1988, as amended by Decree 59 of 1992, ushered in FEPA, which is now a full-fledged Ministry (The Federal Ministry of Environment). The Ministry's goal was to regulate matters and actions concerning the Nigerian environment, its properties, misuse, and the board (Musa and Ikhajiagbe, 2021). Despite the fact that FEPA requires adequate support from the Nigerian government, the level of progress identified by FEPA is not consistent with her stated goals and objectives. This is focused on the fact that the state of our environment is deteriorating at a faster rate than before FEPA was established.

6.7 ASSESSING ENVIRONMENTAL CLEANUP

Looking at the immediate consequences of environmental deterioration in Nigeria shows that it is possible to predict the environment's fate in a couple of years from today. With the steady population increase, this problem is not likely to go away soon unless concerted efforts at remediation are put in place. Each remediation approach has the overarching goal of assisting and maintaining human health and well-being by establishing a long-term solution (Musa, 2019).

It is usually difficult to pin point on a specific type of remediation. This is because it has to be based on different criteria. For example, remediation could be categorized based on site of application, in which case in-situ and ex-situ remediation results. In-situ remediation is carried out on the natural sites of contamination, whereas ex-situ remediation is done away from sites of contamination. The word "natural site" refers to polluted soil that has been treated in its natural environment. Defiled soil that has been transferred or expelled from the subsurface is referred to as ex situ (off site). Containment measures include the construction of engineered barricades, caps, and liners to prevent pollutant movement. Another classification divides remedial steps for polluted media into six categories: isolation and containment, immobilization, solidification, vitrification, soil washing, and soil flushing (GWRTAC, 1997).

Remediation defined based on the force on nature involved in the cleanup can be categorized as physical remediation, chemical remediation, or bioremediation. In physical forms of remediation activities including scooping, physical clearing, booming of water bodies during oil contamination situations, burning, tilling, or physical removal of the contaminant from the site of contamination are some of the measures set in place. These methods don't usually handle the problem effectively as there is always evidence of the contaminant remnants at the sites of contamination. Unless when integrated into other remediation measures, physical remediation may not be effective after all.

The use of chemical agents like surfactants or chemical stimulants implies a chemical remediation. Cleanup crews used over 1.4 million gallons of different chemical dispersants to treat the oil from the Deepwater Horizon oil spill. Typically, such vast quantities are sprayed from an aircraft or helicopter over the open ocean. Chemical dispersants are also sprayed on impacted soils. In a number of cases, the chemical ligating agents like Ethylene Diamine Tetraacetic Acid and Ammonium Oxalate have been used to enhance metal accumulation by plants during remediation (Omoregbee et al., 2017a, b).

Chemical reclamation determines the framework or proportion of material effects on soil in order to improve its properties and increase crop yield. The salts that are harmful to crops are extracted from the soil root layer during this process, resulting in a decrease in the hydrogen and aluminum content of acidic soils. The presence of these components in the soil assimilation complex causes soil debility, physical concoction, and organic features to deteriorate, and soil maturity is damaged. In most soils, the presence of replaceable cations, which may enable the soil to maintain the applied supplement, is minimal. The use of compost has been recently accepted and promoted in aiding soil reclamation, and its sustainability is increased as a result, the production. Recently, an evolving agreement has emerged on the need for both natural and inorganic fertilizer in Sub-Saharan Africa, as continuous primary use of both of these would extensively render the soil more suitable for crop generation.

Chemical reclamation is the method of improving a soil's chemical or physical properties by adding a chemical compound to it in order to increase crop production while minimizing environmental damage. The quantity of large and small nutrients, plant need for mineral salts, soil acidity, soil alkalinity, soil saltiness, and soil microbial action all influence the form of chemicals used. Chemical reclamation can be divided into three categories: salinity-alkalinity regulation, soil structure (condition), and soil acidity regulation. The pH of the soil, the production and quantity of major cations in the soil solution, and the cation exchange in the soil are all affected by chemical substances used to enhance the physico-chemical properties of acidic or alkaline soils. In soil chemical alteration, substances containing the corresponding chemical compounds, such as lime, gypsum, calcium chloride, and sulfuric acid, are used. Chemical materials are used to improve soil structure (i.e. artificial conditioners), enhance soil porosity, improve infiltration characteristics, and increase water penetrability.

During emergency responses to oil spills, chemical dispersants are often used. Consideration should be made on the socioeconomic and environmental consequences of oil spilled pollution with the simultaneous use of chemical degreaser-cleanser for its recovery rather than relying on delayed natural remediation procedures or bio-cleanup.

Chemical land reclamation for agricultural purposes is distinct from that used for water and rubber cleanup. Using a large number of recovering operators, a mechano-chemical cleanup of scrap rubber powder was completed. The recovery process included the use of chemical agents (such as amines, pentachlorothiophenol, and metal chlorides) as well as processing oil as a conditioner. The ability of these reagents is determined by the splitting of sulfur bonds between rubber chains, as well as the climax of the free extreme chain formed as a result of the formation of S-S and C-C bonds. Soil cleaning and soil flushing are two different remediation restructuring methods. Soil washing is an offsite remediation innovation that combines physical and fluid-based separation unit processes to lower waste concentrations and meet site-specific remedial goals, while soil flushing is the removal of pollutants from the soil on site using a proper cleaning procedure. Water is injected into or sprayed against the polluted region, followed by reinjection (USEPA, 2007). Water, an environmental extractant, and a contaminant recoup from the in-situ material are all needed for soil flushing. Despite the fact that this physical and chemical recovery restructuring is optimal, it has unfavorable environmental consequences. For example, the system produces a large amount of solid waste, it is a brief arrangement (where pollutants are present in the area), contaminants can be enacted when soil physicochemical features are altered, and these restyling are typically not "eco-friendly." Furthermore, it is costly, necessitating the use of less expensive cleaning methods.

The use of any biological entity in the reclamation of degraded environments is referred to as bioremediation. During the categorization of bioremediation techniques the type of organism used in the remediation is employed. For example, Phytoremediation refers to the use of plants to achieve this fit. The use of syndicate microorganisms to remove or sequester contaminants from or in polluted media is called microbial bioremediation (Ikhajiagbe et al., 2015, 2017b). Mushrooms and other fungi have been used in remediation, and this is referred to as mycoremediation (Ikhajiagbe, 2010). Vermiremediation employs worms in the remediation procedure. A number of bioremediation interventions have been reported by a number of researchers in Nigeria.

Bioremediation is proving to be a more cost-effective option. Phytoremediation (plants) and rhizoremediation are two biological pathways for remediation (plant and microorganism relationship). Rhizoremediation, the most advanced bioremediation method, combines the removal of specific pollutants from contaminated areas with the fundamental interaction of plant roots and proper microbial flora (Ikhajiagbe, 2016b; Romantschuk, 2000). The majority of bioremediation methods rely on microorganisms to combat contaminants and convert them to non-harmful products. Since bioremediation is only possible in environments that allow for microbial advancement and processes, it is often used in conjunction with environmental restrictions to allow for faster microbial growth and degradation. Bioremediation is a slow and methodical process. A few microorganisms have shown their efficacy as pollutant degraders. Several strains have been found to be active as bioremediation agents, but only under laboratory conditions. The effects of pH, oxygen, temperature, soil structure, appropriate nutrient and dampness measures, poor bioavailability of pollutants, and proximity to other dangerous compounds reveal the need for bacterial development. Although microorganisms can survive in extreme conditions, the majority of them prefer a more favorable atmosphere, which is difficult to achieve outside of the laboratory (Shkidchenko, 2002; Bento et al., 2005). The majority of bioremediation procedures are carried out in a robust state, but anaerobic conditions can also enable microbial degradation of uncooperative particles. For the removal of tough lignin and organo-pollutants, microscopic organisms and growths depend on the synergy of different intracellular and extracellular proteins (Entry et al., 1999; Brakstad and Bonaunet, 2006). Microorganisms have also been observed oxidizing and degrading metal pollutants. Mercury and cadmium can be oxidized, but microorganisms can reduce arsenic and iron. In the presence of sulfur, microorganisms such as *Bacillus subtilis* and sulfate-degrading microscopic organisms may achieve this response (Mulligan et al., 2001). The ease with which overburdening metals and more compounds could be removed using metal-accumulating plants was first noticed in 1983, but the concept has undoubtedly been of great value for as long as three decades on wastewater explosions (Henry, 2000; Chaney et al., 1997).

Bioremediation may not only be restricted to direct use of biological entities to achieve environmental cleanup. In a number of cases, these organisms may have been augmented either through introduction of other secondary organisms which would enhance the remediative capacity of the former. This is referred to as bioaugmentation (Ikhajiagbe and Anoliefo, 2012). During bioaugmentation strategies, microbial inoculants are introduced to soil to enhance plant remediative or tolerance capacities during the remediation (Ikhajiagbe et al., 2013a, b, c, 2014; Chijioke-Osuji and Ikhajiagbe, 2016a). Others may just simply amend the soil with sawdust or other chemical agents to either vary soil pH, water infiltration ability, or soil texture (Ikhajiagbe et al., 2013a, b, c; Anoliefo et al., 2016; Ikhajiagbe, 2016a, b; Ikhajiagbe and Osayamen, 2019). Microbial remediation strategies begin with the assessment of contaminated sites for presence and diversity of resident microorganisms. These usually have capacities for contaminant utilization (Ikhajiagbe and Ogwu, 2020). Evbuomwan et al. (2019) investigated the remediative capacity of crude oil-polluted soil after exposure to poultry manure and phosphate minerals and reported enhanced remediation as a result of the treatments applied. Soil factors have been reported to affect remediation capacities. One of such is cation exchange capacity (Ikhajiagbe et al., 2019).

The use of plants in remediation is almost the most popular type of bioremediation. Plants are efficient at eliminating metals and other pollutants because, like humans, they require specific trace elements to survive. Hyperaccumulators are a group of animals that can accumulate and tolerate high metal concentrations, up to 5% of their dehydrated weight. This characteristic, in combination with size and a rapid growth rate, determines whether an organism is a financially savvy competitor for phytoremediation. The growth and early characteristics of the plants, as well as their high biodiversity, make them ideal for the remediation of areas contaminated by pollutants such as heavy metals.

Heavy metal damage to plants, on the other hand, can cause changes in their metabolic pathways, such as photosynthesis, respiration, and growth, altering plant life setup (Marcelo et al., 2011). The role of bioremediation for the cleanup of environmental and non-biological pollutants, as well as other applications, is depicted as the process of how contaminants interact with the plant system, rhizosphere, and transportation modes in an accurate manner. These structures are interconnected and dependent on plant physiological processes such as photosynthesis, rhizospheric forms, and other forerunners. Therefore, in phytoremediation implementation, distinct methodologies are included depending upon the structured application.

The application of plants in remediation strategies stretches through various kinds of contaminants (Anoliefo et al., 2010; Shittu and Ikhajiagbe, 2013; Ikhajiagbe and Mensah, 2015; Ikhajiagbe and Akendolor, 2016; Ikhajiagbe et al., 2016, 2017a). Ultimately, however, is the capacity for degraded soils to "heal" itself. When left over a long period of time, polluted soils eventually recover due to the activities of physical, chemical, and biological components of the environment. Whereas sunlight increases rate of volatilization of volatile components of contaminants, rainfall enhances wash-off. The soil biota utilizes some of the carbon and eventually the soil is bereft of contaminants. This is what is referred to as natural attenuation or intrinsic bioremediation (Bento et al., 2005; Ikhajiagbe et al., 2013c, 2014; Chijioke-Osuji and Ikhajiagbe, 2016b). Ikhajiagbe and Anoliefo (2017) reported significant reductions in metals and polyaromatic hydrocarbon contents of a spent lubrication oil-polluted soil after exposure to sodium azide and hydroxylamine hydrochloride solutions: implications for intrinsic bioremediation. Omoregie et al. (2020) assessed the quantitative and qualitative basement of microbial presence during phytoremediation of heavy metal polluted soil using *Chromolaena odorata*.

Plant rhizosphere interactions are critical to plant survival in contaminated soils and especially for plants with remediation capacities. Igiebor et al. (2019) reported enhanced growth and development of salinity-exposed rice (*Oryza sativa*) after seedlings were rhizo-inoculated with *Bacillus subtilis* under different pH levels. Similarly, yield assessment of rice (*Oryza sativa*) after rhizo-inoculation with selected plant growth-promoting rhizobacteria in a ferric ultisol was also reported by Ikhajiagbe and Ohanmu (2019).

A couple of other methods of remediation which use an array of technologies to achieve contaminant cleanup abound. Examples are the use of bioreactors as well as sparging. Bioreactors use microorganisms to degrade pollutants in groundwater and soil. Bioreactors, as opposed to natural attenuation and in-situ bioremediation, can prevent and monitor the problems of inadequate indigenous microorganisms and/or low indigenous microbial populations. Air sparging, however, is a subsurface contaminant remediation method in which pressurized air is injected into polluted ground water, allowing hydrocarbons to change state from dissolved to vapor. After that, the air is sent to vacuum extraction systems, which eliminate the pollutants.

Environmental remediation is not a spontaneous operation, and it should not be confused with emergency response following an accident. Following an accident, emergency response typically necessitates immediate action. As a result, a detailed assessment of the situation and formulation of the desired outcome is not only probable but also necessary. A variety of clearly defined measures are often used in the remediation process. The evaluation of the polluted site comes first, followed by the selection of the most suitable remediation strategy. During this time, effective safety measures for the remediation methods chosen are finally implemented. Remediation is finally carried out under the strict supervision of defined protocols. The completion of remediation isn't really complete until the site has been evaluated and deemed remediated. The evaluation of the polluted site comes first, followed by the selection of the most suitable remediation strategy. During this time, effective safety measures for the remediation methods chosen are finally implemented. Remediation is finally carried out under the strict supervision of defined protocols. The completion of remediation isn't really complete until the site has been evaluated and deemed remediated.

The remediation specialist analyzes the site to determine what pollutants are present during Site Assessment. They'll draw a map of the region and take samples of soil and groundwater to determine the level of the contamination. This allows them to estimate how much work would be needed for the remediation. Following that, experts will decide the best cleanup methods. Excavation, dredging, bioaugmentation, soil mixing, oxidation, soil vapor extraction, thermal desorption, and nanoremediation are some of the remediation strategies available, depending on the form and quality of the contaminant. Several fact checks are needed during the remediation process. The concerns that must be considered during the decision-making process range from technological to economic and social factors. For example, the possibility of contaminant exposure to the general public will have to be considered. There's also the matter of occupational exposure as a result of the remediation work – employees would be exposed while the work is being done. Net benefits to the impacted group from the remediation work – remediation can do more good than damage. Wastes are often created during remediation projects, and it is therefore critical that such wastes be safely evacuated to ensure that pollutants do not return to the setting. Costs of remediation work are typically large, necessitating the use of funding mechanisms.

Combining at least two or more remediation methods will increase not only remediation performance but also cost efficiency in practice. However, the cost of remediation, long-term efficacy, socioeconomic effects, commercial availability, wide acceptance, universal applicability over a wide range of pollutants, and applicability to mixed waste are all considerations that could affect the appropriateness and decision of any of the available remedial techniques.

6.8 FUTURISTIC CHALLENGES

Despite the fact that Nigeria is laden with various polluting emitting chemicals, it is fair to consider suitable laws and approaches that are designed to organize and protect human activities related to environmental use. It is becoming increasingly important to ensure that these laws are correctly implemented and that those who break them face consequences. The situation where little attention is paid to the issue of environmental regulations has revealed some slack in the implementation of these norms, implying that anybody can violate environmental laws and get away with it. This proclivity has resulted in environmental blundering on numerous levels. For example, pollution levels arising from vehicles with dreadful exhaust and consuming engines are extremely high, and little has been done to regulate this common occurrence. There are laws in place to protect the environment from pollution and other wastes, but implementation of such laws is important.

Human actions and natural disasters are putting Nigeria's climate in jeopardy. There are already some environmental issues and noticeable scars associated with the degradation of the natural resource base (land, water, and air) on which all life depends. The country's large population of around 200 million people, as well as its 2.8% annual growth rate, coupled with the unguarded human activities, are leading to environmental degradation. Land erosion, deforestation, and land, water, and air pollution are among Nigeria's major environmental concerns. By far the most valuable resource for survival is land. Simply stated, land is the portion of the earth's surface that is not surrounded by water. Wetlands, which are submerged seasonally or indefinitely, must be added to the productive land resources of Nigeria. Most of this land is agricultural, with forest, different forms of trees, and water reservoirs. Future access to rural land for agriculture would be risky if it is left unregulated as it is today. As a result, efforts must be made now to implement land demarcation for rural areas, especially at the Local Government level.

The problem of environmental pollution in the Niger Delta has been exacerbated by the use of fossil fuels, especially oil and gas exploration. Similarly, indiscriminate and illegal tin and columbite mining in the Jos Plateau, as well as other mineral resources in many parts of Nigeria, has badly polluted many areas of the country. It is critical to enact legislation that encourages the growth, production, and use of renewable energy. Despite numerous national attempts to address the problems associated with pollution sources, Nigeria continues to receive a poor rating in terms of environmental performance. According to 2016 Draft of the National Policy on the Environment, the country's Environmental Performance Index (EPI) in 2016 was 58.27, placing it 133rd out of 190 countries surveyed worldwide. Nigeria has a long way to go in terms of achieving the Sustainable Development Goals (SDGs), which are concerned with the conservation and sustainable use of natural resources in order to promote positive socioeconomic development, with climate change exacerbating the situation. Due to its low resilience and adaptive capability, the country is highly vulnerable to severe negative effects of climate change. Nigerian environmental laws cross international and regional agreements/conventions/protocols, as well as national laws, legislation, and bye-laws, since the country is a part of the global environmental management equation. The 1968 African Convention on Conservation of

Nature and Natural Resources, the 1972 UN Conference on the Human Environment (Stockholm Declaration), which established the connection between development and environmental integrity, the 1976 Vancouver Conference on Human Settlements (Habitat I), and the 1985 Vienna Convention on the Protection of the Ozone Layer are just a few of the international agreements/conventions. There are the 1999 Nigerian Constitution, National Policy on Environment, Harmful Wastes (Special Criminal Provisions) Act Cap 165, FEPA Decree No. 58 of 1988 Cap 131, to mention a few.

Distinct objectives and delivery priorities must be developed, as well as new target assessment apparatuses. Industry should be told about effective business operations for a variety of current activities. Industry must play a significant role in educating their employees and hired laborers. Governments and industries should work together to ensure a thorough and unsurprising ecological assessment and surveys of approved destinations, led by appropriately trained and experienced personnel. Infractions should be punished, and the basic framework should be strict. Landowners and administrators, industrial experts, and environmentalists will make informed decisions about reclamation issues by appreciating the need to mitigate the impact of current events. Current and developing practices are constantly evolving and improving, making it essential to stay informed.

6.9 CONCLUSION

Nigeria should enforce extant environmental laws particularly on companies that breach ecological norms in order to remain competitive throughout the period spent on sustainable environmental practices. This is further supported by the fact that the majority of today's emission issues are related to production lines and the manufacturing industry, whether in the small-, medium-, or large-scale businesses. Something must be done to consciously limit the indiscriminate release of contaminant wastes into the environment. Legislation that is already in place must be enforced and strictly followed as a way of maintaining environmental stewardship. Remediation is critical for the restoration of contaminated habitats, but it is certain that if the atmosphere was not polluted ab initio, by properly adhering to procedures and protocols, we would not need to go to the point of remediation; after all, as previously mentioned, nature "heals" itself by natural attenuation. The future is largely dependent on a commitment to properly implement regulations, guidelines, and procedures, not just for remediation but also to ensure a healthy and sustainable environment for all.

REFERENCES

Aboyade, O. (1968). Industrial location and development policy: the Nigerian case. *Nigerian Journal of Economic and Social Studies* 10: 3.

Adebayo, B. (2018). Nigeria overtakes India in extreme poverty ranking [Online]. Accessed 21 December, 2018.

Adegbola, K. (1983). Manufacturing industries. In: Oguntoyinbo J. S. (Eds). *A Geography of Nigerian Development*, 2nd ed., Heinnemann Educational Books Ltd, Ibadan. pp. 326–338.

Adejobi, O. S. and Olorunnimbe, R. O. (2012). Challenges of waste management and climate change in Nigeria: Lagos state metropolis experience. *African Journal of Science and Research* 7 (1): 346–362.

Ahove, M. A. N. (2001). *Environmental Management and Education: An Introduction.* Golden pen Books, Lagos. 219 p.

Ajayi, D. D. (2011). Nigeria's industrial development: issues and challenges in the millennium. Sustainable development and planning V. *WIT Transactions on Ecology and the Environment* 150: 711–723.

Allan, V. K. (1971). Environmental pollution: economics and policy. *American Economic Review, Paper and Proceedings* 61(2): 154.

Amuda, O. S., A. Adebisi, L. A. Jimoda and A. O. Alade (2014). Challenges and possible panacea to the municipal solid waste management in Nigeria. *Journal of Sustainable Development Studies* 6 (1): 64–70.

Anoliefo, G. O., Ikhajiagbe, B., Berena, A. T. and Okoro, R. E. (2010). Bioremediation of crude oil-polluted soil by using *Vernoniaamagdylina* and manure. *International Research Journal of Biotechnology* 1(4): 37–43.

Anoliefo, G. O., Ikhajiagbe, B., Okoye, P. C. and Omoregie, O. (2016). Utilizing local soap-derived biosurfactant for degradation of petroleum hydrocarbon polluted soils: sustainable remediation in focus. *Annals of Science and Technology - B* 1(1): 43–51. Doi: 10.22366/ast.2016.01.007.

Anoliefo, G. O, Ikhajiagbe, B, Omoregie, E. S., Aimiebenomor, R. A. and Okoye, P. C. (2017). Plant growth response and nitrate reductase activities of roots of *Chromolaenaodorata* in a model spent lubricating oil-polluted soil. *StudiaUniversitatisBabeş – Bolyai, Biologia* LXII (2): 97–106. Doi:10.24193/subbbiol.2017.2.09.

Appannagari, R. R. (2017). Environmental pollution causes and consequences: a study. *North Asian International Research Journal of Social and Humanities* 3 (18): 13.

Badejo, M. A. (1998). Agroecological restoration of savannah ecosystems. *Ecological Engineering* 10: 209–219.

Bakare, W. (2020). *Solid Waste Management in Nigeria.* Bioenergy Consult. https://www.bioenergyconsult.com/solid-waste-nigeria/.

Barrow, C. J. (1993). *Developing the Environment: Problems and Management.* U.K. Longman Scientific and Technology, London, 328 p. doi: 10.4324/9781315844923.

Bento F. M., Camargo, F. A. O., Okeke, B. C. and Frankenberger, W. T. (2005). Comparative bioremediation of soils contaminated with diesel oil by natural attenuation, biostimulation and bioaugmentation. *Bioresource Technology* 96: 1049–1055.

Brakstad, O. G. and Bonaunet, K. (2006). Biodegradation of petroleum hydrocarbons in seawater at low temperatures (0–5 degrees C) and bacterial communities associated with degradation. *Biodegradation* 17: 71–82.

Cachada, A., Ferreira da Silva, E., Duarte, A. C. and Pereira, R. (2016). Risk assessment of urban soils contamination: The particular case of polycyclic aromatic hydrocarbons. *Science of The Total Environment* 551–552: 271–284. Doi: 10.1016/j.scitotenv.2016.02.012.

Chaney, R. L., Green, C. E. Filcheva, E. and Brown, S. L. (1994). Effect of iron, manganese, and zinc enriched biosoils compost on uptake of cadmium by lettuce from cadmium contaminated soils. In: Clapp, C. E., Larson, W. E. and Dowdy, R. H. (Eds). *Sewage: Land Utilization and the Environment.* American Society of Agronomy, Madison, WI. pp 205–207.

Dare, A. D. (2007). Recent trends and patterns in Nigeria's industrial development, council for the development of social science research in Africa. *Africa Development* (X)(XII (2): 139–155.

Entry, J., Watrud, L. and Reeves, M., (1999). Accumulation of Cs-137 and Sr-90 from contaminated soil by three grass species inoculated with mycorrhizal fungi. *Environmental Pollution* 104: 449–457.

European Commission. (2006). *Environment Fact Sheet: Industrial Development.* 5 p. Greekwand Publishers. Germany.

Evbuomwan, N. O., Oviasogie, F. E., Oviasogie, P. O. and Ikhajiagbe, B. (2019, October). Remediative capacity of crude oil-polluted soil after exposure to poultry manure and phosphate minerals. *Journal of Applied Sciences and Environmental Management* 23 (10) 1893–1900. Doi: 10.4314/jasem.v23i10.20.

FAO. (2001). Soil Fertility Management in Support of Food security in Sub-Saharan Africa. Available on ftp://ftp.fao.org/agl/agll/docs/foodsec.pdf. 64 p.

FAO. (2015). *Global Forest Resources Assessment 2015.* Food and Agriculture Organization of the UN, Rome, Italy.

GWRTAC. (1997). *Remediation of Metal-contaminated Soils and Groundwater.Tech.* Rep. TE-97-01, GWRTAC, Pittsburgh, PA, GWRTAC-E Series. 61 p.

Hekstra, G. P. (1985). *Productivity and Land Use in Semi-Arid Regions - Lessons from the Sahel. Ecoscript 27.* Foundation for Ecological Development Alternatives (*Stichingmondiaal* Alternative), The Netherlands.

Henry, R. J. (2000). *An Overview of the Phytoremediation of Lead and Mercury.* United States Environmental Protection Agency Office of Solid Waste and Emergency Response Technology Innovation office. Washington, DC, 55 p.

Ifie, J. E., Ifie-Etumah, S. and Ikhajiagbe, B. (2020). Physiological and biochemical responses of selected cowpea (Vignaunguiculata (L.)Walp.) accessions to iron toxicity. *Acta Agriculturae Slovenica* 115(1): 25–38, Ljubljana 2020. Doi: 10.14720/aas.2020.115.1.3.

Igiebor, F. A., Ikhajiagbe, B. and Anoliefo, G. O. (2019). Growth and development of salinity-exposed rice (Oryzasativa) rhizo-inoculated with *Bacillus subtilis* under different pH levels. *Studia Universitatis Babeş-Bolyai Biologia* LXIV(2): 41–53. Doi: 10.24193/subbbiol.2019.2.04.

Ikhajiagbe, B. (2010). *Synergism in Bioremediation: Phytoassessment of Waste Engine Oil Polluted Soils after Amendment and Bioaugmentation.* LAP Lambert Academic Publishing, Köln. 276 p.

Ikhajiagbe, B. (2016a). Possible adaptive growth responses of *Chromolaena* odorata during heavy metal remediation. *IFE Journal of Science* 18(2): 403–411.

Ikhajiagbe, B. (2016b). Biorecovery of a model oil-polluted soil after exposure to solutions of typical salts found in irrigation water. *Studia Universitatis Babes-Bolyai Biologia* LXI(1): 133–146.

Ikhajiagbe, B. and Akendolor, A. (2016). Comparative effects of pretreatment of stem cuttings of*Chromolaena odorata*(Siam weed) with sodium azide and hydroxylamide on the survival and phyoremediative performance in an oil-polluted soil. *Nigeria Journal of Biotechnology* 30(1): 27–39.

Ikhajiagbe, B. and Anoliefo, G. O. (2012). Substrate bioaugmentation of waste engine oil polluted soil. *Research Journal of Environmental and Earth Sciences* 4(1): 60–67. http://maxwellsci.com/jp/abstract.php?jid=RJEES&no=157&abs=10.

Ikhajiagbe, B. and Anoliefo, G. O. (2017). Changes in metals and polyaromatic hydrocarbon contents of a spent lubrication oil-polluted soil after exposure to sodium azide and hydroxlamine hydrochloride solutions: implications for intrinsic bioremediation. *Studia Universitatis Babeş – Bolyai, Biologia* LXII(1): 57–66.

Ikhajiagbe, B. and Musa, S. I. (2020). Application of biosynthesized nanoparticles in the enhancement of growth and yield performances of Rice (Oryza sativa var. Nerica) under salinity conditions in a ferruginous ultisol. FUDMA Journal of Sciences (FJS) 4(1): 120–132.

Ikhajiagbe, B., Anoliefo, G. O. and Imoni, A. E. (2015). Rhizoremediation of poly aromatic hydrocarbon content of a model waste diesel engine oil-polluted soil by some local lawn plant species in Benin City, Nigeria. *Studia Universitatis Babeş-Bolyai Biologia* 60(2): 5–16.

Ikhajiagbe, B., Anoliefo, G. O., Okoh, H. U. and Owenaze, I. (2013c). Impact of organic mulching on the enhanced natural attenuation of a petroleum hydrocarbon polluted soil. *Nigerian Society for Experiential Biology (NISEB) Journal* 13(3&4): 119–130.

Ikhajiagbe, B., Anoliefo, G. O., Omoregbee, O. and Osigbemhe, P. (2014). Changes in the intrinsic qualities of a naturally attenuated waste engine oil polluted soil after exposure to different periods of heat shock. *Resources and Environment* 4(1): 45–53. Doi: 10.5923/j.re.20140401.05.

Ikhajiagbe, B., Anoliefo, G. O., Oshomoh, E. O. and Agbonrienrien, B. (2013a). Effects of watering regimes on the intrinsic qualities of bioremediated waste engine oil-polluted soil. *Annual Review and Research in Biology* 3(2): 107–123.

Ikhajiagbe, B, Anoliefo, G. O., Oshomoh, E. O. and Airhienbuwa, N. (2013b). Changes in heavy metal contents of a waste engine oil polluted soil exposed to soil pH adjustments. *British Biotechnology Journal* 3(2): 158–168. (ISSN: 2231–2927).

Ikhajiagbe, B., Edegbai, B. O. and Aibuedefe, F. (2017b). Hydrocarbon utilization by culturable microbial species in the rhizosphere of *Eleusine*indica in oil-polluted soils. *Nigeria Society of Experimental Biology, NISEB Journal* 17(1): 1–8.

Ikhajiagbe, B., Edegbai, B. O., Omoregie, G. O. and Eweka, A. M. (2017a). Assessment of the phytoreclamation of an oil-contaminated soil cultivated with *Cynodon dactylon*, *Eleusine indica*, and *Eragrostis tenela*. *Studia Universitatis Babeş – Bolyai, Biologia* LXII (1): 43–55.

Ikhajiagbe, B. and Musa, S. I. (2020). Application of biosynthesized nanoparticles in the enhancement of growth and yield performances of rice (*Oryza sativa* var. Nerica) under salinity conditions in a ferruginous ultisol. *FUDMA Journal of Sciences* 4(1): 120–132.

Ikhajiagbe, B., Musa, S. I., Aliu, E. and Ogunro, J. (2020b). Studies of foliar bioconcentration of metals by *Vernonia amygdalina* in a model heavy metal-polluted soil. *Studia Universitatis "Vasile Goldiş", Seria Ştiinţele Vieţii* 30(2): 64–68.

Ikhajiagbe, B, Musa, S. I. and Okeme, J. O. (2019). Effects of changes in soil cation exchange capacity on the reclamation of lead by Eleusine indica (L.) Gaertn *FUDMA Journal of Science* 3(4): 176–183.

Ikhajiagbe, B. and Mensah, J. K. (2015). Phytoscreening of some local ornamental plants in Benin City, Nigeria for remediation of polyaromatic hydrocarbons in waste engine oil-polluted soil. *Nigerian Annals of Natural Sciences* 15(1): 066–071. (Published by Fac. of Natural Sci., Ambrose Alli University, Ekpoma, Nigeria).

Ikhajiagbe, B. and Ogwu, M. C. (2020). Hazard quotient, microbial diversity, and plant composition of spent crude oil-polluted soil. *Beni-Suef University Journal of Basic and Applied Sciences* 9: 26. Doi: 10.1186/s43088-020-00052-0.

Ikhajiagbe, B., Ogwu, M. C., Fawehinmi, F. O. and Adekunle, I. J. (2020a). Comparative growth responses of *Amaranthus* spp. L. in humus and ferruginous ultisols using plant growth promoting rhizobacteria (*Pseudomonas* species). *South African Journal of Botany* 137 (2021): 10–18. Doi: 10.1016/j.sajb.2020.09.029.

Ikhajiagbe, B. and Ohanmu, E. O. (2019). Growth and yield assessment of rice (*Oryza sativa*) after rhizo-inoculation with selected plant growth-promoting rhizobacteria in a ferric ultisol. *Studia Universitatis "Vasile Goldiş", Seria Ştiinţele Vieţii* 29(3): 134–143.

Ikhajiagbe, B. and Osayamen, K. I. (2019). Application of silver nanoparticles during bioremediation of petroleum hydrocarbon-polluted soil by Eleusine indica – inhibition or improvement? *Research Journal of Chemical Science* 9(3): 7–16.

Ikhajiagbe, B., Shittu, H. O. and Imhagbe, P. (2016). Bioaccumulation of heavy metals by *Vernonia amygdalina* in a waste engine oil-contaminated soil. *University of Benin Journal of Science and Technology* 4(1): 94–103.

Ingo, W. (1972). Environmental control and patterns of international trade and investment: an emerging policy issue. *Banca Nationale del lavoro Quarterly Review* 100: 83.

Marcelo, P. G. and Teresa, C. L. (2011). Ecophysiological and anatomical changes due to uptake and accumulation of heavy metal in *Brachiaria decumbens. Scientia Agricola* 12: 685 p.

Millennium Ecosystem Assessment. 2005. http://www.millenniumassessment.org/en/index.aspx.

Mulligan, C. N., Yong, R. N. and Gibbs, B. F. (2001). Remediation technologies for metal-contaminated soils and ground water: an evaluation. *Engineering Geology* 60: 193–207.

Musa S. (2019) Isolation and identification of diesel oil-degrading bacteria in used engine oil contaminated soil. *Journal of Applied Science and Environmental Management* 23(3): 431-435.

Musa, S. I. and Ikhajiagbe, B. (2020). Assessment of physico-chemical properties of ferruginous ultisol in Benin City, Edo State – Possible impact on plant distribution. *Studia Universitatis "Vasile Goldiş", Seria Ştiinţele Vieţii* 30(2): 88–95.

Musa S., Ikhajiagbe B. (2021) The growth response of rice (Oryza sativa L. var. FARO 44) in vitro after inoculation with bacterial isolates from a typical ferruginous ultisol. Bulletin of the National Research Centre 45(20): 1–20. Doi: 10.1186/s42269-021-00528-8.

NEST. (1991). *Nigerian Threatened Environment: A National Profile.* National Environmental Study/Action Team, Ibadan, 288 p.

Oladipo, E. O. (2008a). *Managing the Nigerian Environment for Sustainable Development.* Draft Report for the Federal Ministry of Environment, Housing and Urban Development. 121 p.

Oladipo, E. O. (2008b). Climate change and sustainable livelihood: greening options for Nigeria. *Paper Presented at the First National Summit on the Environment*, Abuja, Nigeria. 121 p.

Omoregbee, O., Anoliefo, G. O. and Ikhajiagbe, B (2017a). *Chelator-Assisted Remediation of Aluminium by Vernonia Amygdalina Del. Plant Morphological, Antioxidant and Microbiological Response to Soil Contamination by Aluminium.* LAP Lambert Academic Publishing, Köln, 128 p.

Omoregbee, O., Ikhajiagbe, B., Igiebor, F. A., Chukwu, V. and Anoliefo, G. O. (2017b). Effect of ethylene diamine tetraacetic acid and ammonium oxalate on the prevalence of microorganisms and removal of aluminum in soil by bitter leaf plant (Vernonia amygdalina Delile). *Journal of Applied Science and Environmental Management* 21 (4): 648–653.

Omoregie, G. O, Ogofure, A. G., Ikhajiagbe, B. and Anoliefo, G. O. (2020). Quantitative and qualitative basement of microbial presence during phytoremediation of heavy metal polluted soil using *Chromolaena odorata. Ovidius University Annals of Chemistry* 31(2): 145–151. Doi: 10.2478/auoc–2020-0023.

Onyemelukwe, J. O. C. (1983). Structural and locational characteristics of manufacturing. In: Oguntoyinbo, J. S., Areola, O. O. and Filani. M. (eds), *A Geography of Nigeria Development.* 2nd ed. Heinemann Educational Books, Nigeria Limited, Ibadan, Nigeria, pp. 296–310.

Oyebode, O. J. (2018). Evaluation of municipal solid waste management for improved public health and environment in Nigeria. *European Journal of Advances in Engineering and Technology* 5 (8): 525–534.

Paul, R. E. and John, P. H. (1971). Impact of population growth. *Science* 171 (3977): 1213.

Romantschuk. (2000). Effect of inoculation of a TOL plasmid containing mycorrhizosphere bacterium on development of Scots pine seedlings, their mycorrhizosphere and the microbial flora in m-toluate-amended soil. *FEMS Microbiology Ecology* 31:127–141.

Shittu, H. O. and Ikhajiagbe, B. (2013). Phytoaccumulation of heavy metals in an oil-spiked soil by *Vernoniaamygdalina* after exposure to sodium azide solutions. *Nigeria Journal of Life Sciences* 3 (2): 169–179.

Shkidchenko. (2002). Selection of microorganisms capable of degrading petroleum and its products at low temperatures. *Applied Biochemistry and Microbiology* 38: 437–440.

Sinton, H. M. (2011). Native prairie reclamation, Alberta environment. http://www.albertapcf.org/rsu_docs/pcf_recl-update_final_110917.pdf, 6 p.

Sridhar, M. K. C., Ana, G. R. E. E. and Laniyan, T. A. (2019). Impact of sand mining and sea reclamation on the environment and socioeconomic activities of Ikate and Ilubirin coastal low income communities in Lagos metropolis, Southwestern Nigeria. *Journal of Geoscience and Environment Protection* 7: 190–205. Doi: 10.4236/gep.2019.72013.

Stephen, E. (2013). Humans: the real threat to life on Earth. In: *Ten Billion*. The Guardian, International Edition, Guardian News and Media Limited or its affiliated companies (2020), 8 pp.

Tomori, S. (1977). *Environmental Pollution in a Rapidly Developing Economy: The Nigerian Experience*. Department of Economics. University of Lagos, Lagos, 6 p.

USEPA. (2007). *Treatment Technologies for Site Clean-up: Annual Status Report*. 12th ed. Tech. Rep. EPA-542-R-07-012, Solid Waste and Emergency Response (5203P), Washington, DC.

WHO. (2017). *Guidelines for Drinking-Water Quality: Fourth Edition Incorporating the First Addendum*. World Health Organization, Geneva. Licence: CC BY-NC-SA 3.0 IGO. 631 p.

7 Sustainable Wastewater Treatment Using Microalgae Technology

Emeka G. Nwoba
Murdoch University

John N. Idenyi
Ebonyi State University

Christiana N. Ogbonna and James C. Ogbonna
University of Nigeria

Mathias A. Chia
Ahmadu Bello University

CONTENTS

DOI: 10.1201/9781003180289-7

7.1 INTRODUCTION

The human population growth and the corresponding increase in the demand for food, energy, and bio-based resources have intensified urban, agricultural, and industrial activities. This intensification has resulted in the production of large volumes of wastewater. Sometimes, this wastewater is partially treated, and in other cases not treated at all (Nwoba et al. 2020e) and are directly discharged into the environment (including water bodies), causing eutrophication and ecosystem disturbances due to excessive nutrient supply. For instance, according to Food and Agricultural Organization, about 60% of wastewater generated anthropogenically from municipalities globally is discharged without treatments to coastal marine waters, rivers, streams, and estuaries (Neveux et al. 2018). The contaminants in the wastewater, particularly phosphorus, nitrogen, metals, and hydrocarbons, build up in these water bodies and lead to eutrophication and destruction of ecological communities. These profound negative impacts of wastewater on the environment and aquatic ecosystems have been recognized as a significant problem of global proportion (Smith 2003). Consequently, mitigation technologies such as bioremediation have attracted a great deal of attention with the focus on reducing the negative impacts of the pollution on the environment.

In bioremediation, living organisms are used to break down, remove, neutralize, and recover contaminants which make it an attractive treatment process from the environmental viewpoint. During this process, naturally occurring biological organisms (via biostimulation) or exogenous organisms (via bioaugmentation) can be utilized to break down or absorb a broad range of pollutants. In this context, bioremediation employs inexpensive and low-technology methods that can be conducted in situ (on site) or ex situ (off site) and, generally, have a wide public appeal (Vidali 2001). The efficiency of this process strongly depends on the inherent metabolic activity of the biological agents of interest, with the environment meant to provide the resources needed for the optimum growth of the organisms. Interestingly, the targeted pollutants usually constitute these resources and must

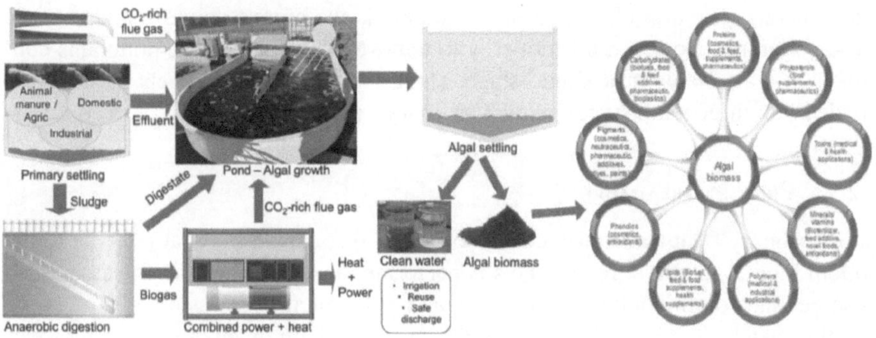

FIGURE 7.1 Scheme of microalgae-based wastewater treatment coupling the production of bioproducts. (Modified from Nwoba et al. 2020e.)

be the limiting nutrient source (e.g. nitrogen, phosphorus, iron, organic matter, petroleum hydrocarbons) under favourable environmental conditions. Innovative and sustainable wastewater treatment techniques are highly sought after in order to make wastewater remediation economical and produce marketable bioproducts instead of waste.

Microalgae-based wastewater treatment technologies represent scalable techniques for meeting the new demand of improved wastewater remediation that is low energy consuming and produces microalgal biomass that can be valorized to different bioproducts (Koutra et al. 2018). Significant merits of using microalgae in wastewater treatments are that effluents are cleaned up through nutrients removal, and the resultant biomass provides bioproducts such as biofuel, feed, biofertilizer, and biomaterials (Figure 7.1). A wide range of microalgal cultivation systems are suitable for algal-based wastewater treatment including classical raceway ponds, high rate algal ponds, tanks, and closed photobioreactors, depending on the purpose (Nwoba et al. 2020e). Overall, microalgae have the potential to assimilate pollutants in wastewater and recycle nutrients into useful bioproducts (Figure 7.1).

7.2 BASIC WASTEWATER TREATMENT PROCESSES AND TECHNOLOGIES

Water is indispensable for life on earth. It is used for various human activities which results in large volumes of wastewaters. These wastewaters contain a lot of both organic and inorganic pollutants which, if disposed untreated, results in various types of environmental problems. Wastewaters are generated from homes, industries, commercial establishments such as banks and markets, and agricultural facilities. It is therefore very important to treat these wastewaters either for recycling (reuse) or to avoid the detrimental effects of discharging untreated wastewater on the ecosystem. Wastewaters contain many substances in high concentrations depending on the source. These pollutants include faecal matter, detergents, herbicides, pebbles, pieces of wood, glasses, rags, plastics, pharmaceuticals, oils and fats, carbohydrates (polysaccharides, disaccharides, and monosaccharides), nitrogen, phosphorus, as well as many metals. The aim of wastewater treatment is therefore to remove and degrade these pollutants. However, it is not easy to remove all or most of the pollutants contained in wastewater in a single stage. Hence, wastewater treatment is therefore divided into three main stages: namely primary, secondary and tertiary treatments.

7.2.1 Primary Treatment

In the primary treatment stage, coarse materials that can easily float or settle to the bottom by gravity are removed. This is performed through physical processes such as screening, comminution, grit removal, and sedimentation. Screening involves the use of screen (a device used to hold solids present in the influent of wastewater treatment plants) to remove materials that could damage the plant or reduce its efficiency. The size of screen to be used depends on the size of the suspended solids. There are various kinds of screens used for physical removal of large debris from wastewater, namely coarse, fine, and micro screens. Coarse screens have large openings that range in size from 6 to 150mm and are used to remove large objects such as rags, small logs of wood, foams, etc. This step is followed by grit (sand and broken glasses) removal. The last solid to be removed under primary treatment is the fine particles via gravity sedimentation. During this step, influent is allowed to flow into large sedimentation/clarification tanks, which vary in shapes and sizes (diameter of 3–90m for circular and 15–100×15–24m for square tanks). This is left to stay for some hours to enable the suspended materials settle at the bottom or rise to the top to form scum. The sludge and scum are subjected to secondary treatment, while the clarified effluent is subjected to tertiary treatment.

7.2.2 Secondary Wastewater Treatment

The effluent from primary treatment is subjected to secondary treatment. The aim of secondary treatment is to use physical phase to remove settleable materials and biological process to remove suspended and dissolved organic compounds. After the secondary treatment, the water has some degree of effluent quality and is called secondary treated water. During secondary treatment, dissolved and colloidal compounds whose concentrations are determined by a measure of their biochemical oxygen demand are removed by the activities of microorganisms such as bacteria and protozoa that degrade biodegradable and soluble organic pollutants like sugars, fats and short-chain carbon compounds from various sources. Waste stabilization pond is the most widely used facility for secondary treatment of wastewaters.

Wastewater treatment is carried out in wastewater treatment plants in most developed societies. However, not every country has well-planned wastewater treatment plants because of the construction and maintenance costs (Nwoba et al. 2020e). The simplest basic wastewater treatment outfit, especially in developing countries, is waste stabilization pond. These are special kind of ponds designed for the purpose of treating wastewater. Inside the ponds, indigenous microorganisms act on the organic matters, and some pathogens in the wastewater reduce their contents by using them as food for energy and growth, and they generate carbon dioxide and water from such activities. The products of the microbial activities are more stable than the original organic matter content in wastewater. Therefore, such ponds are called stabilization ponds. Because the microorganisms that degrade the organic matter contents in wastewater require oxygen for respiration, if the wastewaters are discharged into water bodies untreated, such waste degrading microorganisms reduce the oxygen contents in the water bodies leading to suffocation and death of aquatic life in such water. Wastewater enters the pond as influent from one side and after residing there for a stipulated time is channelled into another pond as effluent for further treatment or discharged into the environment such as natural water bodies like lakes or rivers or used for irrigation. The ponds may be built in series so that different stages of the treatments are done in each pond. Several microorganisms live in the stabilization pond and act synergistically or in succession on the organic matter contents of the wastewater. Since the organisms are varied, they belong to different trophic levels. The organisms at the lowest trophic level metabolize the utilizable part of the organic matter in the wastewater converting them to cell biomass, by-products, and products. These biomass, products, and by-products are useful nutrients for organisms in the next trophic level, and this goes on in succession based on the contents of the wastewater. Bacteria are the most important microorganisms in the primary wastewater stabilization pond since they use most of the organic matter in the pond, but they consume oxygen thereby requiring the pond to be aerated. Another very important and unique group of pond biome are microalgae. Microalgae are photosynthetic and thus synthesize their own organic matter using nutrients in the wastewater and by-product from bacterial activities such as CO_2. Microalgae are very important as a component of the stabilization pond, since through their photosynthetic activities, they generate the much-needed oxygen for the respiration of other aerobic life forms in the pond. Another source of oxygen in the pond is from the atmosphere and this dissolves in the liquid surface and creates aerobic liquid surfaces.

Most of the pathogenic bacteria and viruses are killed or deactivated by the shift in the pH of the pond due to the activities of microalgae coupled with the direct effects of ultraviolet light that reach the pond surfaces. As the stabilization ponds mature, most of the protozoan cysts and oocytes in the water including the eggs of helminths settle to the bottom of the pond by the force of gravity. The sediment is discharged as sludge which requires further treatment to decontaminate the sludge.

7.2.3 Tertiary Treatment

After the secondary treatment, the effluent is passed on for tertiary treatment which is also known as advanced treatment. At this stage of treatment, there is further reduction in the turbidity of the effluent from secondary treatment. In particular, organic compounds, nitrogen, phosphorus, metallic pollutants, and even pathogens from the secondarily treated effluents are removed. The effluent from the tertiary treatment plants is expected to be clean enough to be discharged to the environment. The effluents are usually fit for use in irrigating food crops, golf courses, and even for drinking. The conventional tertiary treatment methods include coagulation, filtration, use of activated carbon to adsorb organic maters, reverse osmosis (Kimura et al. 2004), and disinfection (Gerba and Pepper 2019). Other advanced treatments include oxidation using ozone, ultraviolet radiation, and chlorination (Dhodapkar and Gandhi 2019).

7.2.4 Microalgae Cultivation As an Advanced Wastewater Treatment Option

All these methods listed above are not very efficient for nutrient removal. Besides, most of them are very technical and expensive. However, many strains of microalgae are known to be very efficient in removing these pollutants. After the primary and secondary treatment, the effluents though apparently clear and clean still have high concentrations of some compounds such as inorganic nitrogen and phosphorus, and some micronutrients that cause eutrophication and other long-term pollution when discharged into the environments. Such partially treated wastewaters are ideal growth medium for most microalgae. Microalgae offer tertiary wastewater treatment by utilizing the inorganic nitrogen and phosphorus in wastewater for growth and production of useful metabolites. Microalgae are used in tertiary wastewater treatments by growing in primarily treated wastewater to remove $N-NH_4^+$, $P-PO_4^{3-}$, $N-NO_3^-$, and chemical oxygen demand (COD). The cultivation time can be as long as 30 days, but a high percentage nutrient removal rate can be achieved within 8 days (Arias et al. 2018). One major advantage of wastewater treatment using microalgae is that the biomass harvested after growth in wastewater can be used for bioenergy production (Arias et al. 2018) and many other useful metabolites (Vadiveloo et al. 2019).

7.3 WASTEWATER SOURCES AND THEIR NUTRIENT PROFILES

The impact of anthropogenic activities on natural resources is attracting increasing interests globally. Algal biotechnology promises solution to this challenge with a special focus on the growth physiology of algae, which is crucial to algae overall performance (Doria et al. 2012). The ability of microalgae to assimilate inorganic and organic compounds makes them perfect organisms for wastewater treatment. These nutrients (both macro and micro) largely meet the biochemical needs of microalgae. The macro-nutrients include sodium, calcium, potassium, magnesium, chloride, sulphur, phosphate, ammonium, and bicarbonate. In contrast, micronutrients are also known as trace elements since they are mostly required in little amounts. They comprise manganese, sulphur, zinc, copper, and cobalt (Nwoba et al. 2020e; Abdel-Raouf et al. 2012a). These nutrients are synthetic chemicals and are supplied externally. Many of these nutrients are found in wastewaters. Therefore, various microalgae when grown in wastewater absorb and assimilate these nutrients (contaminants) resulting in water purification (Abdel-Raouf et al. 2012a). Besides, these applications are eco-friendly, and desirable biomass can be generated for further applications in pharmaceuticals, fine chemicals, and biofuels production among others (Doria et al. 2012). Thus, this emerging technology is critical to environmental and public health protections by preserving the quality of water bodies (Nwoba et al. 2020e).

In efforts to avoid relying on exorbitant conventional chemicals, cheap wastewater as a sustainable alternative source of nutrients is being proposed for algae farming. A variety of wastewaters are utilized by different types of algae as growth media, which perform last-stage or tertiary treatment via the uptake of nitrogen, phosphorus, and other elements present in wastewater. Furthermore, this technique has the capacity to yield high cell concentrations, which could be utilized as a precursor for the production of algal-based valuable products (Miao and Wu 2006; Bouarab et al. 2004). Reports have emerged showing that wastewater as a nutrient source could yield sufficient biomass for large-scale algal biofuel production (Cai et al. 2013c).

The first step in the application of microalgae in wastewater treatment is to look for and collect species with desirable characteristics. The selected microalgal species must be robust on wastewater as growth media, demonstrate rapid growth, and proficient in nutrient uptake from the wastewater among other characteristics (Abou-Shanab et al. 2013). Investigations revealed the application of different groups of microalgae in wastewater treatment for the assimilation of pollutants. The microalgal species frequently used to treat pollutants (nitrogen and phosphorus, chemical/biological oxygen demand) from a variety of wastewaters are *Botryococcus* species, *Scenedesmus* species, and *Chlorella* species as detailed below.

7.3.1 Municipal Wastewater

Municipal/domestic wastewater is generally derived from home organic matters or sewage and surface water pollution (Lundquist et al. 2010). Its nutrient profile could vary significantly from one location to another because of disparities in sources and seasons, and its qualities demonstrate relatively similar attributes as a result of everyday human activities (Lundquist et al. 2010). Freely available municipal wastewater has been observed to contain rich nutrients (primarily carbon, nitrogen, and phosphorus) that are suitable for the growth and biochemical activities of algae (Table 7.1). Nevertheless, domestic wastewater also holds substantial quantities of potentially toxic metals such as copper, lead, and zinc, which in excess might negatively influence the performance of algae (González et al. 1997). The use of domestic

TABLE 7.1
Comparison of Nutrient Removal Efficiency by Different Microalgal Species Cultivated in Municipal/Domestic Wastewater

Microalgae Species	Source of Wastewater	Parameter (mg/L)	Removal Efficiency (%)	Ref.
Chlorella sp.	Municipal/ Domestic	COD (231), TN (40.65), NH_4^+ (33.4), PO_4^{3-} (5.65)	COD (50.9), TN (68.4), NH_4^+ (82.4), PO_4^{3-} (83.2)	Wang et al. (2010)
Chlorella sp.	Municipal wastewater	NO_3^- (0.87-0.01), PO_4^{3-} (6.68-02.94)	NO_3^- (88.5), PO_4^{3-} (55.9)	Idenyi et al. (2019)
Chlorella minutissima	Domestic wastewater	BOD (99.5-4.9), COD (149.7-14.9), NO_3^- (3.6-0.3), NH_4^+ (39.5-3.9), PO_4^{3-} (4.47-1.15)	BOD (95), COD (90), NO_3^- (91.6), NH_4^+ (90), PO_4^{3-} (74.3)	Sharma and Khan (2013)
Botryococcus braunii	Greywater collected from hostel	BOD (279.6–66.7), COD (986-85.6), NH_4^+ (14.2-0.3), NO_3^- (0.24-0.07), TP (9.6-2.2)	BOD (76.1), COD (91.3), NH_4^+ (97.8), NO_3^- (69.6), TP (97.6)	Gokulan et al. (2013)

wastewater for the growth of microalgae has been widely investigated with records of huge success for nutrient removal (Table 7.1). For instance, Li et al. (2011a) proved that domestic wastewater supplied nutrients for the cultivation of *Chlorella kessleri* and *C. protothecoides* and resulted in the production of maximum biomass of 2.01 and 1.31 g/L, respectively. Also, these two freshwater microalgae species showed the potentials to grow in wastewater supplemented with synthetic nutrients.

7.3.2 Animal Manure/Agricultural Wastewater

Animal manure/agricultural wastewaters are generated from livestock and agricultural practices. Examples are wastewaters from dairy cattle, swine, poultry, and beef cattle production. These wastewaters contain high concentrations of nitrogen and phosphorus, making them good sources of nutrients for microalgal growth and sustainable algal diesel generation (Ahluwalia and Goyal 2007). Supplementing agricultural wastewater with other essential nutrients is strongly encouraged to guarantee the attainment of the Redfield ratio requirement with respect to the nitrogen and phosphorus ratio. This clearly will support the sustainable growth and productivity of microalgae. The enlargement of the broad base of animal manure practice could result in the availability of more nutrients concentrations leading to more economic algal farming (Abou-Shanab et al. 2013). The use of microalgae to remove nutrients from agricultural wastewater has been well documented (Table 7.2).

TABLE 7.2
Comparison of Nutrient Removal Efficiency by Different Microalgae Species Cultivated in Animal Manure/Agricultural Wastewater

Microalgae Species	Source of Wastewater	Parameter (mg/L)	Removal Efficiency (%)	Ref.
Chlorella sp.	Aquaculture	TN (130-11.7), NH_4^+ (15-0), TP (71.3-5)	TN (91), NH_4^+ (100), TP (93)	Udom et al. (2013)
Chlorella vulgaris	Treated piggery wastewater	TN (56), TP (13.5)	TN (49), TP (18)	Abou-Shanab et al. (2013)
Chlorella zofingiensis	Piggery wastewater	COD (3460-697.5), TN (139-24.1)	COD (79.8), TN (82.7)	Zhu et al. (2013)
Scenedesmus obliquus	Treated piggery wastewater	TN (1280–599), NH_4^+ (1197-520), NO_3^- (6.8-1.1), TP (4.3-0.7)	TN (53), NH_4^+ (57), NO_3^- (84), TP (83)	Jimenez-Perez et al. (2004)
Chlorella pyrenoidosa	Dairy effluent after algae bead treatment	BOD (1200-146), COD (2900-448), NH_4^+ (35-0.6), PO_4^{3-} (30-0.87)	BOD (88), COD (85), NH_4^+ (98), PO_4^{3-} (97)	Yadavalli and Heggers (2013)

BOD, biochemical oxygen demand; COD, chemical oxygen demand; NH_4^+, ammonium; NO_3^-, nitrate; PO_4^{3-}, phosphate; TN, total nitrogen; TP, total phosphorus.

TABLE 7.3
Comparison of Nutrient Removal Efficiency by Different Microalgae Species Cultivated in Industrial Wastewater

Microalgae Species	Sources of Wastewater	Removal Efficiency (%)	Ref.
Chlorella vulgaris	Leather industry	BOD (22), COD (38), TKN (73), NO_2^- (89), NO_3^- (91.5), TP (94), NH_4^+ (80)	Rao et al. (2011)
Chlorella vulgaris	Textile industry	COD (62.3), NH_4^+ (45.1), PO_4^{3-} (33.3)	Lim et al. (2010)
Chlorella sakina	Tannery industry	NO_3 (9.1). NH_4^+ (62.0), PO_4^{3-} (81.9)	Jaysudha and Sampathkumar (2014)

BOD, biochemical oxygen demand; COD, chemical oxygen demand; NO_3^-, nitrate; NH_4^+, ammonium; PO_4^{3-}, phosphate; TKN, total Kjeldahl nitrogen; TN, total nitrogen; TP, total phosphorus.

7.3.3 Industrial Wastewater

Industrial wastewater is the effluent or liquid waste resulting from the activities of production industries. The nutrient contents of these wastewaters differ significantly from one industry to another according to the raw materials used. A specific treatment method is needed for the different industries since each industry might contain a particular mixture of nutrients or pollutants (Lim et al. 2010). Previous studies suggest that industrial wastewaters contain somewhat more potentially toxic metals and less of the crucial nutrients like nitrogen and phosphorus in contrast to both manure and domestic wastewaters. However, several industrial effluents have been treated using microalgae (Table 7.3).

7.3.4 Wastewaters from Oil and Gas Industries

Wastewater is usually obtained from onshore and offshore wells during the mining of oil and gas. They are originally found within the sedimentary rock rich earth's crust, beneath the earth surface. The wastewaters from oil and natural gas mining are carried in fluids generated by hydraulic force on hydrocarbon deposits and flow through the artificial channel to the earth's surface. Water and some chemicals are sometimes pumped into the production region to enhance the operational safety and recovery speed of these products (Neff et al. 2011). The produced complex mixture comprised of dissolved organic and inorganic chemicals that demonstrate nutrients ability for the growth of algae. These nutrients include inorganic salts, metals, radioisotopes, and other forms of organic compounds, mainly hydrocarbons (Nwoba et al. 2020e). Environmental concerns linked to the chemical composition of the produced oil and natural gas can be addressed by treating produced water with microalgae (Graham et al. 2017).

TABLE 7.4
Comparison of Nutrient Removal Efficiency by Different Microalgae Species Cultivated in POME Wastewater

Microalgae Species	Sources of the Wastewater	Removal Efficiency %	Ref.
Chlamydomonas sp.	POME from facultative pond	COD (8.6), TN (73), NH_4^+ (100), TP (63.5)	Ding et al. (2016)
Scenedesmus dimorphus	POME	BOD (71.5), COD (79), TN (87.5), NH_4^+ (88.5), TP (92.5)	Rajkumara and Takriffab (2015)
Chlamydomonas incerta	POME	COD (3.8), NO_3^- (12.5), NH_3^+ (7.5), PO_4^{3-} (70)	Kamyab et al. (2017)
Spirulina sp.	POME from anaerobic fourth pond	COD (50), TN (40), TP (40)	Soetrisnanto and Hadiyanto (2014)
Spirulina platensis	POME	BOD (78.3), COD (84.9), TN (91), NH_4-N (93.8), TP (96.8)	Rajkumara and Takriffab (2015)
Characium sp.	POME from anaerobic pond	COD (45.4), TN (88.6), NH_3 (90.4), NH_4^+ (87), TP (99.5), PO_4^{3-} (99.1)	Tamil Selvam et al. (2015)

BOD, biochemical oxygen demand; COD, chemical oxygen demand; NH_4^+, ammonium; NO_3^-, nitrate; PO_4^{3-}, phosphate; TN, total nitrogen; TP, total phosphorus.

7.3.5 Palm Oil Mill Effluent (POME)

POME is a wastewater generated from the palm oil producing industry. High concentration of degradable organic materials is found in raw or treated POME. The compositions of POME are water (95%–96%), total solid (4%–5%), suspended solid (2%–4%), and oil (0.6%–0.7%) (Muttamara 1996). Water bodies can be polluted by untreated wastewater (e.g. POME) due to oxygen attenuation and other associated effects. The oil content of POME is either suspended in supernatant or suspended on the upper layer of the mixture. Microalgae are applied nowadays to remove contaminants from POME which in turn generate useful products. The main aim of wastewater treatment is the security of the environment and the public health by protecting the water bodies which provide important water resources. Table 7.4 shows the nutrient composition of POME and contaminants removal competence of different microalgal species in POME.

7.4 FACTORS THAT INFLUENCE MICROALGAL NUTRIENT RECOVERY IN WASTEWATER

Elevated nutrient content in most aquatic environments is one of the key resultants of the disposal of wastewater; this often leads to favourable conditions for undesirable phytoplankton blooms. Studies have established the potency and efficiency of microalgae in removing toxic metals, nitrogen, and phosphorus from wastewater in a controlled environment (Abdel-Raouf et al. 2012b; Cai et al. 2013b). The propensity of microalgae to assimilate nutrients has made them excellent for the tertiary treatment/remediation of wastewater. Spikes in the pH of the surrounding microenvironment due to algae growth induces phosphorus precipitation and stripping of ammonia to the surrounding air (Larsdotter 2006). Also, there has been a progressive increase in chemical-based pharmaceutical contaminants (PCs), textile dyes, nanoparticles, and other emerging environmental contaminants in effluents discharged into most freshwater bodies across the globe. This is fast becoming an emerging environmental concern because of the considerable eco-toxicity and the potential associated health implications which might arise from the use of these waters (Xiong et al. 2018). Bioadsorption/remediation of these contaminants ranging from commonly used analgesics like diclofenac, ibuprofen, paracetamol to hormonal drugs like progesterone and norgestrel via the controlled use of microalgal cells has been well reported (de Wilt et al. 2016; Peng et al. 2014).

The enzymatic and biochemical pathways of nutrient adsorption and conversion largely depend on external and microenvironmental conditions that microalgal cells are subjected to. These factors significantly affect remediation performance with regard to nutrients removal proficiency and biomass productivity of microalgae. Some of the conditions that affect the biochemical pathway of algal nutrient adsorption include but are not limited to dissolved inorganic carbon (such as CO_2), phosphorous, nitrogen source, N/P and C/N ratios, pollutants with cationic groups, light intensity, temperature, and nature of wastewater (Hongyang et al. 2011; Larsdotter 2006; Li et al. 2019; Richmond and Hu 2013; Xiong et al. 2018).

7.4.1 Utilization of Dissolved Inorganic Carbons

Cyanobacteria and algae have been established to utilize CO_2 and HCO_3^- for photosynthesis among the microbial phototrophs due to the presence of carbonic anhydrase (CA) on their cell surface (Aizawa and Miyachi 1986). Most microalgal species mainly take up CO_2 during photosynthesis. Kinetic studies reveal that most of the CO_2 utilized by cyanobacteria follow the ionic conversion of HCO_3^- to CO_2 mediated by CA located on the cell surface. This implies that the actual molecular species which cross the selective plasmalemma is mostly free CO_2. There is thus an apparent variation in the manner of utilization of dissolved inorganic carbon (DIC) for photosynthesis in a consortium of microalgae used for bioremediation. An elevated rate of DIC utilization under low CO_2 directly depends on both CA and a CO_2 concentrating mechanism (Aizawa and Miyachi 1986). Exogenous CO_2 promotes autotrophic growth efficiency of microalgae because they play a role in controlling water pH during photosynthesis (Hu et al. 2012; Mehrabadi et al. 2015; Min et al. 2012). pH and its varying levels during photosynthesis can potentially cause a significant influence on nutrients removal (Park and Craggs 2010). Total nitrogen and phosphorus removal can be increased from 78% to 93% when the pH of consortia of green microalgae such as *Chlorella* spp. with bacteria and single algae system is regulated around the circumneutral zone compared to unregulated systems with more than two times lower nutrient uptake rates (Liang et al. 2013). Also, the consortia of algae and bacteria combined with pH regulation provide a more efficient means of nutrients removal than unialgal systems. CO_2 addition, as demonstrated by studies, has shown that it can have great impacts on wastewater treatment by microalgae (Hu et al. 2012; Li et al. 2011b).

7.4.2 Phosphorous, Nitrogen source, N/P and C/N Ratios

Phosphorus is an essential macro-nutrient for algal growth; it is assimilated in ionic form, orthophosphate (PO_4^{3-}). The enzymatic conversion of orthophosphates to organic phosphates is mediated by phosphatases at the outer cell membrane; this primarily occurs when inorganic phosphates are limited. Microalgae are thus able to take up phosphorus in excess and store within the cells in the form of polyphosphate granules (Larsdotter 2006). Stored polyphosphate is adequate for prolonged algal growth under deprivation of phosphorus. The growth rate of an alga or its consortium may therefore not instantaneously react to alterations in the surrounding concentration of phosphorus, in contrast to the immediate responses to temperature and light. Nitrogen (10% of algal biomass) is the second most important nutrient for microalgae growth (Becker 1994; Larsdotter 2006). The most predominant nitrogen compounds assimilated by microalgae are ammonium (NH_4^+) and nitrate (NO_3^-) (Oliver and Ganf 2002). Amino acids such as arginine, glutamine, and asparagine can be taken up by cyanobacteria. Also, some species have been established to fix nitrogen gas (N_2) (Bhaya et al. 2000). When ammonium is available, no alternative nitrogen sources will be assimilated because nitrogen fixation from sources other than ammonia is energy demanding and only occurs in some cyanobacteria when no other nitrogen compounds are available in sufficient amounts (Benemann 1979). Thus, the remediation efficiency of most microalgae largely depends on the nitrogen source.

N/P and C/N ratios which have been established to play an important role in nutrient assimilation (Richmond 2008) vary significantly with regard to the nutrient content in wastewater. Carbon would be the limiting factor when microalgae can use organic carbon through mixotrophic or heterotrophic metabolism in wastewater treatment. When the C/N ratio is inadequate for reproduction, microalgae can obtain CO_2 through aeration. Thus, the balance of C/N ratio becomes an efficient, simple, and cost-effective method of promoting nitrogen assimilation by microalgae (Zheng et al. 2018). The N/P ratio varies significantly across wastewater source and type. There is evidence that N/P ratio and algal biomass productivity are negatively correlated. Biomass productivity decreases when external N/P ratio increases (Mehrabadi et al. 2015; Ruiz-Martínez et al. 2015). N/P ratio above the optimal level might trigger phosphorus limitation while at levels below the optimal would result in nitrogen limitation (Olguín 2012; Wang et al. 2010). Nitrogen and phosphorus can limit the growth of microalgae, and their abundance can significantly influence both their structure and bioremediation potential (Barsanti and Gualtieri 2014).

7.4.3 Pollutants with Cationic Groups

The most effective way through which microalgae eliminate organic contaminants from an aqueous phase environment is via biodegradation. Microalgae can transform complex molecules into simpler molecules by catalytically degrading their complex parent compounds (Peng et al. 2014). Products of this degradation are mainly extracellular polymeric substances (proteins, polysaccharides, enzymes, substituents, polysaccharide-linked methyl and acetyl groups, and lipids; Figure 7.2) (Khona et al. 2016). Extracellular polymeric substances (EPS) are capable of aggregating

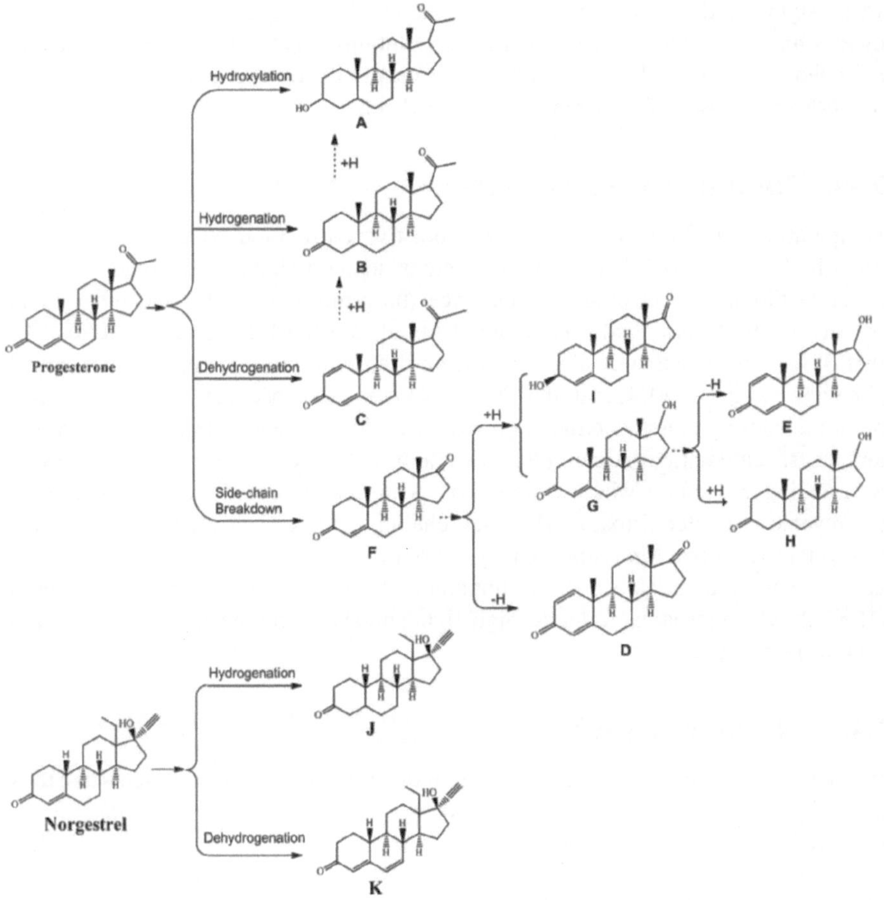

FIGURE 7.2 Tentatively proposed biotransformation pathways for progesterone and norgestrel by *Scenedesmus obliquus* and *Chlorella pyrenoidosa*. Products A–K represent the transformation products. A = 3b-Hydroxy-5a-pregnan-20-one; B = 3,20-Allopregnanedione; C = 1,4-Pregnadiene-3,20-dione; D = Androsta-1,4-diene-3,17-dione (ADD); E = 17b-Boldenone; F = 4-Androstene-3,17-dione (AED); G = Testosterone; H = 5a-Dihydrotestosterone (5a-DHT); I = Epi-androsterone; J = 4,5-Dihydronorgestrel; K = 6,7-Dehydronorgestrel. (Peng et al. 2014.)

into a matrix of hydrated biofilms and act as external digestive system that keeps extracellular enzymes close to the cells. This facilitates the metabolism of organic compounds, which occurs in either dissolved, colloidal, or solid form. The efficiency of this process is affected by the nature of pollutants in wastewater. The microalgal cell wall is an assemblage of polymers with salient similarities to several carbohydrates (cellulose, pectins, hemicelluloses, arabinogalactan proteins, extensin, and lignin) with negatively charged dominant functional groups, such as carboxyl, phosphoryl, and amine. These groups confer a negative charge on the cell wall of microalgae (Xiong et al. 2018). Since adsorption is an extracellular process, sorption varies significantly according to the hydrophobicity, structure, and functional groups of PCs and other contaminants. Pollutants containing cationic groups are actively attracted to the microalgal surface via electrostatic interaction and, resultantly, lead to ineffective biosorption by remediating algal cells.

7.4.4 Temperature and Light Intensity

Temperature and light intensity have a positive correlation with regard to algal growth (Li et al. 2011b). Higher light intensities combined with exogenous CO_2 promote biomass production and enhance the removal of COD and nitrogen in the treatment of municipal wastewater (Li et al. 2012). Elevated temperatures like that of the tropics enhance algal growth until an optimum temperature is reached (Chevalier et al. 2000; Larsdotter 2006). However, a subsequent increase in temperature leads to a rapid decline in growth rate which implies a decrease in sorption and, thus, remediation. Algal cultures in humid climates are always at the risk of overheating especially where evaporation is inhibited (Larsdotter 2006). This is the reverse in cooler climates where the challenge rather is growth limitation due to low temperatures. Microalgae easily get photo-inhibited by high light intensities at low temperatures. However, at temperatures around optimum for algal growth (15°C–25°C), microalgae tolerate high light intensities before growth retardation sets in (Borowitzka 1998).

7.4.5 Nature of Wastewater

Wastewaters are direct product/cocktail of industrial and domestic effluents and thus are unique in their composition (physical and chemical properties). Temperature, pH, nutrient levels, and toxic substances as discussed greatly influence bioremediation and adsorption rate, algal growth, nutrient recovery, and biomass productivity (Larsdotter 2006; Li et al. 2019; Xiong et al. 2018). Pollutants are primarily described by their nutrient contents and nutrient strength. As discussed, algae have their optimal cultivation conditions, and thus, in practical situations, the wastewater nutrients are rarely in synchrony with the optimal nutrients profile for algae growth. Consolidating for this mismatch may pose several challenges that require the use of strategies to ensure that bioadsorption and remediation occur at an optimal rate.

One of the major challenges in microalgal bio-removal/remediation research is the complexity in developing generic technologies because a large number of variables are involved. These variables influence and determine the selection process for biomass processing due to the complex characteristics of wastewater. Trial tests of cultivating algae in wastewater are mostly at the laboratory scale. Growing algae on an industrial scale to remediate wastewater continues to be problematic. Further research via in-depth studies through the use of cutting-edge molecular approaches, such as metabolomics, metagenomics, metaproteomics, and metatranscriptomics to reveal algal community and species-specific interactions that are most suited for scale-up, should be comprehensively carried out to optimize operating parameters.

7.5 MICROALGAE CULTIVATION SYSTEMS AND MODES SUITABLE FOR WASTEWATER TREATMENT

Most of the industrial effluents contain a lot of both organic and inorganic pollutants that need to be removed or degraded before discharge into the environments. Many species of microalgae are known to be able to remove these pollutants through various processes such as absorption (Chen et al. 2012), adsorption, and degradation (Carpenter 1987), while others can de-toxify them, making them safe for discharge into the environment. Since the resulting algae biomass have many applications while other species are able to produce many extracellular metabolites from these wastewaters, a lot of interest has been focussed on valorization of these wastes by microalgae, a process whereby the wastes are treated by converting the pollutants to either microalgae biomass and/or extracellular products. The primary effluents often contain, in addition to metabolizable nutrients, some suspended solids that make the effluents very turbid and unsuitable for the growth of microalgae. However, some primary and secondary treatments are used to remove these suspended solids and, in some cases, hydrolyse some polymer materials into monomers that can easily be metabolized by microalgae. These primary and secondary treatment processes such as anaerobic digestion also reduce the organic load and turbidity of the effluent thereby improving light penetration which is very crucial for photoautotrophic growth of microalgae. However, efficient wastewater treatment by microalgae requires efficient culture systems which are optimized for the various modes of nutrition by various species of microalgae.

7.5.1 CULTIVATION SYSTEMS

Wastewater treatment by microalgae can be done in either open air photobioreactors or closed photobioreactors. The open air photobioreactors include natural water bodies such as lagoons, lakes, and ponds and artificial ponds such as raceway and circular ponds. There are many types of closed photobioreactors, but the ones that have been used for wastewater treatment include bubble column photobioreactors, tubular photobioreactors, and flat plate photobioreactors. Bubble column photobioreactors are similar to the bubble column heterotrophic bioreactors except that they are made of transparent materials for light penetration and their diameters are usually very small (less than 10 cm) for increased illuminated surface to volume ratio. Mixing is achieved by air bubbling through spargers of different designs. The major design criteria for bubble column photobioreactors are the diameter of the column, the height to diameter ratio, and the nature of sparger, while the operation parameters are mainly the aeration rate (which determines the mass transfer characteristics) and light intensity. Although bubble column photobioreactors have been used for cultivation of various species of microalgae, their application in wastewater treatment is limited by the difficulty in scaling them up (Alhamed et al. 2014).

Tubular photobioreactors consist of transparent tubes (glass, Perspex, polyvinylchloride, etc.) that can be horizontal, inclined, or spirally wound around a central support (Nwoba et al. 2016). The diameters of the tubes vary, but typically they are less than 10 cm in diameter. They are usually equipped with pumps for circulation of the cells, but some are air lift systems. They have gas exchange unit for addition of CO_2 and stripping of oxygen. They can also be equipped with heat exchangers for cooling or heating of the culture depending on the weather condition. A major problem with tubular photobioreactors is growth of the algae on the inner surface of the tubes, which can be solved by using foam bulb in washing. Tubular photobioreactors have been shown to be very efficient in cultivation of various species of microalgae (Nwoba et al. 2016), but the difficulty in scaling them up limits their application for wastewater treatment.

As the name implies, flat plate photobioreactors are rectangular box with transparent surface made of transparent material such as glass, plexiglass, and polycarbonate. They can be illuminated indoor or installed outdoor for solar light illumination. The optimum depth (light path) depends on the target strain and the expected light intensity, but the aim is usually to maximize the illuminated surface to volume ratio. Mixing is by bubbling of air through a long sparger fixed at the bottom of the plate. In some designs, broth circulation is assisted by means of a pump. It can be vertical or inclined at different angles for optimal illumination (Nwoba et al. 2020d). Depending on the diameter, cleaning is easier, while oxygen build-up is lower compared to tubular photobioreactors, but surface growth can still be a problem, depending on the strain and the mixing system. The main design parameters are the width:diameter:height ratio and the type of sparger and temperature control system, while the operation parameters include the orientation, inclination angle, sparging (which determines the gas holdup, mixing efficiency, and mass transfer), hydraulic retention time, and operation temperature. Like other closed photobioreactors, they have been extensively used for cultivation of many strains of microalgae (Sun et al. 2019), but their application for wastewater has been limited by the problem of scale.

The advantages and disadvantages of closed photobioreactors depend on the type (Table 7.5), but on the whole they have higher mass transfers, higher illumination surface to volume ratios, better light distribution, lower risk of contamination, and generally higher volumetric productivities. However, they are too expensive for wastewater treatment, and it is technically difficult to scale them up to the required scales for wastewater treatment. Although all of them have been used to cultivate many strains of microalgae using different types of wastewater (Nwoba et al. 2016, 2019a, 2020e; Chuka-ogwude et al. 2020), the main purpose of such cultures is valorization (production of useful metabolites) rather than large-scale wastewater treatment. Most of the municipal and industrial wastewater effluents are produced in hundreds of tons per day which are too much to be treated in closed photobioreactors. Moreover, some of the secondary effluents are either too turbid or toxic so that they are usually diluted, thus increasing the total volumes to be treated.

Open air photobioreactors seem to be more suitable for large-scale wastewater treatment by microalgae. These include open tanks, natural bodies of waters, and artificial ponds. Open tanks are of various shapes, but cylindrical ponds made of plastics, galvanized steel, or concrete are common. They are manually mixed at time intervals and used mainly for on-site cultivation of microalgae for aquaculture. However, they can also be used for on-site treatment of the effluent which contains high concentrations of nitrogen, phosphorus, and some carbon resulting from the

TABLE 7.5
Some Photobioreactor Systems Used for Wastewater Treatment

Production System	Advantages	Disadvantages
Natural bodies of water	Very cheap, low energy cost (only for harvesting), utilizes already existing water bodies so there is no need for arable land, employs already adapted mixed strains of microalgae	Destruction of ecosystems, poor mass transfer, low cell growth and nutrient removal, restricted to places with natural water bodies.
Artificial Ponds (Raceway, circular, etc.)	Cheap to construct and operate, easy to clean and maintain, low energy costs	Low productivity, requires large areas of land, they are used for only strains that grow under selective conditions (pH, salinity, etc.), poor mixing, poor utilization of carbon dioxide.
Columns	Small area of land is required, spacing and column height can be adjusted to suit the required light illumination, good mixing and high mass transfer, low O_2 build-up	Low illuminated surface to volume ratio, relatively expensive depending on the mixing system.
Tubular (horizontal, inclined, spirally wound)	Tube diameter can be adjusted to meet the required illuminated surface to volume ratio, high productivity, relatively cheap, can be horizontal, inclined at various angles or spirally wound to suit the available space and required treatment rates	Difficult to scale up; there are often wall growth, difficult to clean; horizontal tubes require large area of land; if the tubes are long, there are gradients in pH, dissolved carbon dioxide and oxygen concentrations along the tube.
Flat plates (inclined, vertical)	Can be inclined at various angles or vertical, low build-up of dissolved oxygen, easier to clean than tubular photobioreactors, the illuminated surface to volume ratio can be adjusted to suit the strain and type of wastewater, high cell growth and nutrient removal rates	Difficult to scale up, depending on the mixing system, wall growth can be a problem.

fish excretion and residual feed. Natural bodies of water include lakes and lagoons. They are usually mixed naturally by wind convection and support the growth of naturally selected indigenous microbial consortium composed of various species of microalgae, bacteria, and fungi in various proportions. They are usually operated as a continuous system with variable hydraulic retention time which typically depends on the rate of effluent discharge. The major advantages include low cost and high stability of the microbial community. They are very good for large-scale wastewater treatment. The major disadvantages are that the microbial growth rate and thus the treatment efficiency are very low and depend a lot on the type of wastewater and the rate of discharge. They are also restricted to areas with such natural bodies of water, and separation of biomass from the effluent is very expensive because of the low-standing biomass concentration and the heterogeneity of the microbial species.

Various types of artificial ponds such as circular ponds, rectangular ponds, and raceway ponds have been extensively investigated for cultivation of various species of microalgae and for wastewater treatment. Circular ponds were very popular in China and many Asian countries, but raceway ponds are now the most widely used algae ponds all over the world. The cost of construction depends on many parameters, but it is considered to be one of the cheapest algae culture systems. Large-scale raceway ponds range from 1,000 to 5,000 m^2. They are usually built on compacted soil underlined with polymers for culture retention (Craggs et al. 2012). Important construction parameters include the channel length, width, depth, lining, as well as whether completely open or inside green house. They are usually built in units of about 100 m^2 and the depths varying from 25 to 50 cm, depending on the species and location. The operating parameters are the working volume, the standing biomass concentration, the hydraulic retention time, supply of carbon dioxide, and speed of paddlewheels. For good light penetration, standing biomass concentration is controlled at about 0.5 g/L, but higher biomass concentrations favour better light absorption, cheaper cost of harvesting the algae, and lower risk of contamination. Maximum performance is achieved by ensuring adequate illumination, carbon dioxide supply, good mass transfer (to avoid oxygen build-up), temperature control (which affects cell growth and culture evaporation), and prevention of dust pollution (which affects the turbidity and thus light penetration) (Rayen et al. 2019).

High rate algal ponds are basically open air algae ponds operated in such a way to achieve very high rates of nutrients (C, N, P) removal while at the same time significantly reducing pathogen load in wastewater. It is currently one of the most widely used microalgae wastewater treatment system (Park et al. 2011). The high rates of nutrients removal are possible because they are usually shallow (~25 cm) to allow good light penetration and efficiently mixed by using paddle wheels. However, in order to maximize carbon dioxide transfer, the ponds are equipped with a sump of about 1.5 m into which carbon dioxide is bubbled. They are usually operated as a continuous system, and the dilution rates depend on the standing biomass concentration, the concentrations of the pollutants in the inflow wastewater, and the weather condition (temperature and solar light intensity). Improvement in the mixing efficiency in ponds has been achieved by modification of the conventional paddle wheels. Zeng et al. (2016) reported that a novel paddle wheel with 15° inclined angle of the blades resulted in significant improvement in the mixing characteristics of the raceway pond. Pathogen removal is also one of the key factors in wastewater treatment. This is partly achieved by solar light radiation which is only achieved in very shallow ponds. For deeper ponds, the radiation reaching the bottom of the pond is not enough to kill the pathogens, but when inclined planes were installed in the ponds, the pathogens were exposed to strong solar irradiation, leading to increased death rates of the pathogen (Hawley and Fallowfield 2018).

On the whole, the open systems for wastewater treatment have advantages of being simple, cheap to construct and operate, and easy to scale up. However, the disadvantages include microbial contamination (though algae can work in synergy with bacteria for increased efficiency of wastewater treatment), predation, high rates of evaporation rates, inefficient light distribution, relatively poor mixing, requirement for large area of land, and inefficient CO_2 absorption (Wang et al. 2013).

Immobilized microalgae can also be used for treatment of wastewater. The immobilization can be in matrix or as biofilms. Matrix (immobilization carriers) used for immobilization of microalgae includes k-carrageenan, alginate, polymer foam, hollow fibre, and chitosan (Sawayama et al. 1998). Immobilization is used to achieve high rates of treatment due to higher cell concentrations, and more importantly it helps in separating the cells after treatment. In treatment systems with suspended algae cultures, separation of the cells after treatment is both economically and technically challenging. If the cells are not separated, they will lyse and release the accumulated nutrients back to the environment. The cost of separating the cells can represent a significant percentage of the treatment cost. However, with cell immobilization, it is very easy to separate and recycle the cells. One of the great limitations of immobilized microalgae for wastewater treatment is the difficulty in scaling up. Thus, batch process is limited to only very small facilities. However, continuous process can be easily operated at large scale, the dilution rate depending on the immobilized cell concentration, the concentration of the pollutants in the feed stream, and the desired contaminant concentrations in the effluent. Light supply is a major problem for cells entrapped in matrix. Thus such immobilized cells are better for heterotrophic cultures.

Biofilm algae culture is another system that is used for wastewater (Kesaano and Sims 2014). It is a type of immobilized system whereby consortia of photosynthetic unicellular and multicellular microalgae as well as other heterotrophic microorganisms such as bacteria and fungi grow together on solid surfaces. They are simple, energy efficient, and cheap to establish. Furthermore, separation of algae biomass from the bulk of wastewater and dewatering of the harvested biomass are easier and cheaper (Miranda et al. 2017).

Membrane bioreactor is a system where the microalgae are separated from the wastewater by a semi-permeable membrane through which the nutrients diffuse from the wastewater to the cells due to concentration gradient. This is also a type of immobilized cell system. However, it is used mainly at laboratory level since it is very expensive and technically challenging to construct a large-scale system. Membrane fouling and membrane stability are also major problems with this type of system. The main advantages of membrane bioreactors include possibility to achieve high cell concentration, and thus high rate of nutrient removal, protection of the cells from inhibitory effects of high-molecular-weight inhibitory substances in the wastewater, and more importantly, easy separation of pure microalgae biomass for various applications.

7.5.2 Modes of Nutrition

Most wastewaters contain a lot of organic carbons (glucose, fructose, disaccharides, polysaccharides, glycerol, organic acids, etc.). The nature and concentrations of these organic carbons depend on the type of wastewater as well as the method of primary and secondary treatments employed. Fortunately, many species of microalgae are able to metabolize a wide range of these organic carbons (Ogbonna et al. 2000). In terms of mode of nutrient, most of the pollutants (different forms of nitrogen, phosphorus, metals, and many other pollutants) can be metabolized or removed by microalgae under photoautotrophic conditions. In photoautotrophic culture, microalgae grow by the process of photosynthesis using sunlight as the source of energy and inorganic carbon (bicarbonate or CO_2) as the source of carbon. In the process, they also use different forms of nitrogen and phosphorus for various metabolic activities. In this mode of nutrition, they can also absorb (bioaccumulation) or adsorb some metals and other contaminants. However, cell growth rates and standing biomass concentrations under photoautotrophic conditions are usually very low, leading to very low rate of pollutants removal. This is especially so in turbid wastewaters which make light penetration into the culture very poor. However, many strains of microalgae are also heterotrophic, that is, they can grow in dark using organic carbons as the sources of both carbon and energy. Thus, they can be used to treat organic carbon-containing wastewater in dark, using the conventional heterotrophic bioreactors. This system is not common because microalgae may not be as efficient as other heterotrophic bacteria in dark. Furthermore, accumulation of some of the useful metabolites by microalgae is comparatively low in dark (Ogbonna et al. 1997, 1998, 1999). Thus, for the purposes of valorization of the wastes, light illumination is important.

Since most of the microalgae culture systems for treatment of organic carbon-containing wastewaters are illuminated, the mode of nutrition is mixotrophic or photoheterotrophic rather than photoautotrophic (Ogbonna and McHenry 2015; Ogbonna and Moheimani 2015). In mixotrophic mode of nutrition, microalgae use light and organic carbon as the sources of energy and both organic and inorganic carbons as the sources of carbon. On the other hand, photoheterotrophic mode of nutrition involves the use of light as the energy source while organic carbon is used as the carbon source. In practical term, mixotrophic and photoheterotrophic cultures can be used interchangeably since both involve the presence of light and organic and inorganic carbon sources. Nevertheless, in mixotrophic cultures, photoautotrophic and heterotrophic metabolic activities proceed independently, depending on the species (Ogbonna et al. 2002a, b), while in photoheterotrophic cultures, light is required for metabolism of the organic carbon. Depending on microalgae species and culture conditions, light can inhibit organic carbon uptake, while under other conditions, the presence of organic carbon depresses photosynthetic O_2 evolution and inhibits respiration and enzymes of Calvin cycle (Liu et al. 2009). However, in most cases, cell growth in mixotrophic cultures is much higher than in photoautotrophic and heterotrophic cultures (Ogbonna et al. 2002a, b). Higher cell growth in mixotrophic cultures can be attributed to the fact that organic carbon assimilation is not limited by lack of oxygen since it is continuously generated by photosynthesis, while oxygen concentration does not rise to inhibitory level since it is continuously being used for organic carbon assimilation. Furthermore, organic carbon assimilation (respiration) results in carbon dioxide evolution which stimulates photosynthesis. Better pH stability has also been reported for mixotrophic cultures (Kong et al. 2011).

It is also important to note that outdoor microalgae cultures for treatment of organic carbon-containing wastewater involve a cyclic mode of nutrition – heterotrophic growth at night and mixotrophic growth during the day. This results in continuous cells growth and nutrient removal both during the day and night.

7.6 ALGAL BIOFUEL PRODUCTION

The growing demand for energy, expansion of global economy, emission of greenhouse gases, and depletion of fossil fuel reserves are global challenges that have necessitated a change to renewable energy sources and biofuels (Milano et al. 2016). Biofuels production from first-generation and second-generation raw materials, and non-edible crops are at the moment not feasible due to competition for agricultural land and other sustainability challenges (Sims et al. 2010). In contrast, microalgae are currently viewed as promising feedstocks for bioenergy production owing to their natural ability to drive lipid-rich biomass production through the process of photosynthesis. The type of biofuel that can be produced from microalgal biomass depends on its macromolecular composition, and examples of microalgal biofuels are biodiesel via transesterification, bioethanol/biobutanol via fermentation, biohydrogen via photobiological reaction, and biogas via anaerobic digestion, of microalgal biomass (Ishika et al. 2017). Among these bioenergy options, biodiesel remains the most studied and feasible option when wastewater is utilized for microalgal growth.

7.6.1 Biodiesel

Microalgae are known to accumulate large quantities of lipids, hence, a suitable feedstock for the production of biodiesel. The fatty acid (FA) profile of microalgal lipids is comparable to those of animal fats and vegetable oils (Koutra et al. 2018). The physical and chemical properties of the biodiesel are strongly dependent on FA chain length, level of saturation and unsaturation, and the degree of branching of the FA components (Yu et al. 2017; Shin et al. 2015). Aliphatic FAs that have 16 and 18 carbon atoms and high percentage of saturation and mono-unsaturation (e.g. oleic acid, C18) are attractive for premium quality biodiesel (Shin et al. 2015). Notably, algal biodiesel is typified by high percentage of linolenic acid (18:3n3) (Chisti 2007). Based on the European Biodiesel Standards EN14214, high percentage of linolenic acid is not desirable as it promotes oxidation; thus, C18:3 in biodiesel should not be more than 12% of total FAs (Shin et al. 2015). Palmitic (C16:0), stearic (C18:0), oleic (C18:1), linoleic (C18:2), and linolenic (C18:3) acids are the most abundant FAs in microalgae (Knothe 2009). High-value FAs such as eicosapentaenoic (C20:5) acid (EPA) and docosahexaenoic (C22:6) acid (DHA) occur in minute amounts in majority of microalgal species (Table 7.6). Nevertheless, certain species of diatoms and eustigmatophytes can accumulate EPA and DHA in incredible amounts depending on culture conditions (Huerlimann et al. 2010). Furthermore, dinoflagellates and haptophytes are natural producers of EPA and DHA (Brown et al. 1996).

While dozens of researchers cultivated microalgae in wastewater and analysed the potential of the biomass produced for biodiesel production based on lipid content and FA profiles, only a few of these studies looked at pivotal parameters such as kinematic viscosity (fuel quality during storage), cetane number (ignition delay time), free FAs (high value impact biodiesel quality and quantity), and iodine value (level of unsaturation) that affect the quality of biodiesel (Cai et al. 2013a; Kim et al. 2015). To be suitable for biodiesel application, FAs should have a small number of double bonds to avoid oxidative instability of the final product. High percentage of saturated

TABLE 7.6
Potential Bioproducts from Microalgae Grown in Wastewater

Product	Examples	Application	Microalgal Sources
Lipids	Hydrocarbons (Botryococcene), Triglycerides, Polyunsaturated fatty acids (eicosapentaenoic acid, docosahexaenoic acid, ϒ-linolenic acid, arachidonic acid)	Biofuels, nutritional/food supplements, feed additives, infant formula	*Botryococcus braunii*, *Chlorella vulgaris*, *Nannochloropsis salina*, *Phaeodactylum tricornutum*, *Scenedesmus* sp., *Arthrospira* sp., *Schizochytrium mangrovei*
Pigments	Chlorophyll a, chlorophyll b, C-phycocyanin, β-carotene, phycoerythrin, lutein, astaxanthin, allophycocyanin	Natural food colourant, pigmentation, cosmetics, food additive, feed additive, pharmaceutical, fluorescent agent	*Chlorella vulgaris*, *Chlorella sorokiniana*, *Dunalella salina*, *Haematococcus pluvialis*, *Arthrospira platensis*, *Aphanizomenon flos-aquae*, *Porphyridium cruentum*
Minerals and vitamins	Magnesium, Selenium, Calcium, Phosphate, Iodine, Vitamins B1, B2, B6, B12, C, E, folic acid, biotin	Food supplement, feed supplements, biofertilizers	*Euglena gracilis*, *Dunaliella tertiolecta*, *Isochrysis galbana*, *Arthrospira platensis*, *Nannochloropsis oculate*, *Tetraselmis suecica*
Phytosterols	Crinosterol, stigmasterol, fucosterol, dinosterol, isofucosterol	Food additives, pharmaceuticals	*Phaeodactylum tricornutum*, *Nannochloropsis gaditana*, *Isochrysis galbana*, *Dunaliella tertiolecta*, *Tetraselmis suecica*, *Pavlova lutheri*
Polymers	Polyhydroxy alkanoates, exopolysaccharides	Industrial and medical uses	*Chlorella vulgaris*, *Scenedesmus acuminatus*, *Arthrospira platensis*, *Synechocystis* sp., *Porphydrium* sp., *Nostoc muscorum*
Amino acids	Leucine, glycine, aspartic acid, glutamine, alanine, lysine, arginine, valine, cysteine, asparagines	Food supplements, feed additive, nutraceuticals	*Euglena gracilis*, *Anthrospira platensis*, *Nannochloropsis oculata*, *Porphyridium cruentum*, *Rhodomonas salina*
Phenols	Gallic, ρ-coumaric, ferulic, caffeic, salicylic acids	Cosmetics	*Phaeodactylum tricornutum*, *Dunaliella salina*, *Nannochloropsis salina*, *Desmodesmus communis*, *Chlorella vulgaris*
Phycotoxins	Okadaic acid, ciguatoxin, domoic acid, gambieric acid, gonyautoxins, yessotoxins	Health and pharmaceutical applications	*Nitzchia purgens*, *Gambierdiscus toxicus*, *Alexandrium lusitanicum*, *Amphidinium* sp., *Dinophysis* sp.
Mycosporine-like amino acids	Sporopollenin, Scytonemin, Shinorine	Sunscreen, UV-screening agent	*Aphanizomenon flos-aquae*, *Chlorella fusca*, *Chlorella minutissima*, *Scytonema* sp.

FAs is characteristic of high-quality biodiesel as it boosts the cetane number (Islam et al. 2013; Gopinath et al. 2009). High value of cetane number indicates short ignition delay time. Free FAs content influence the oxidative stability of biodiesel and optimal ratio of C16:1, C18:1, C14:0 is 5:4:1 (Schenk et al. 2008). The density and kinematic viscosity of biodiesel are affected by the percentage of unsaturated FAs. C18:3 FA should not surpass 12% and FAs with ≥ 4 double bonds should not exceed 1% in biodiesel (Moser 2014). Generally, optimal balance of saturated and unsaturated FAs must be ensured to achieve high-quality biodiesel.

For microalgae cultivated in wastewater, the lipid accumulation and FA profile are influenced by a number of factors: (1) growth conditions such as pH, temperature, CO_2 addition, inoculum size, (2) nitrogen availability, (3) microalgal strains, (4) microalgae-bacteria interactions, (5) wastewater characteristics and history, (6) exponential-stationary growth phases, and (7) batch, fed-batch, continuous, and semi-continuous cultivation modes (Vadiveloo et al. 2019). Lipid contents of microalgae vary with species, and a range between 4.5% and 80% dry weight of biomass has been reported (Hu et al. 2008). Under nitrogen starvation, the lipid contents dramatically increase in all microalgae, with the exception of cyanobacteria (Hu et al. 2008). However, nutrient stress stagnates microalgal growth and lockdown photosynthesis, routing energy to triacylglycerols accumulation (Ajjawi et al. 2017). Lipid yield is usually low at exponential growth phase with increase in lipid contents as the cells approach late exponential and stationary phases (Hu et al. 2008; Xu et al. 2008).

7.6.2 Biogas

Whole or spent microalgal biomass can be subjected to anaerobic digestion to produce biogas (biomethane), which is composed of 55%–75% methane and 25%–45% CO_2 (Zamalloa et al. 2011). During anaerobic digestion, microorganisms naturally degrade algal biomass in the absence of oxygen to generate methane at a volume $> 6.0 \times 10^8$ tons/year (Kwietniewska and Tys 2014). Other biogas mixtures such as biohydrogen and biohythane (5%–25% methane) are available renewable gases that can be produced from microalgal biomass. Spent algal biomass from oil extraction or hydrogen production is good as an anaerobic digestion feedstock as they lack lignin (Shuba and Kifle 2018). The methane can be combusted to generate electricity or even used as a fuel gas source. The sludge from the anaerobic digestion process can be used as biofertilizer. Several factors influence the economics and effectiveness of biogas production from microalgae through anaerobic digestion. These factors include pretreatment of biomass, strain of microalgae, anaerobic digestion conditions, and microalgae growth methods. Towards a circular bioeconomy, effluents from the anaerobic digestion process can be used as a growth medium for microalgae culture. The biomass generated from this culture can be fed back to the anaerobic digestion facility to produce biogas, which is utilized as an energy source. Some microalgae such as *Tetraselmis* sp., *Chlorella* sp., and *Chroococcus* sp. have been studied in wastewater environment with the aim of producing biogas (Prajapati et al. 2014). The carbon-nitrogen, C:N, ratio of the wastewater tends to influence the quantity of the biogas produced due to impact on the chemical constituent of the final biomass (Ward et al. 2014). Low C:N ratio of effluents promote ammonia accumulation and increase in pH, which adversely affect biogas production through inhibition of methanogenesis (Ward et al. 2014). Notably, the final biogas produced usually has impurities such as hydrogen sulphide, nitrogen, hydrogen, and at least 40% CO_2. These impurities require to be removed to increase the calorific value of the biogas, and this process is called biogas upgrading. Microalgae have high affinity for carbon sequestration, and this feature has been exploited in the upgrading of biogas by photosynthetic capture of the CO_2 for biomass formation (Ward et al. 2014).

7.6.3 Bioethanol

Besides biogas production from microalgal biomass, bioethanol can also be produced via fermentation of the carbohydrate content by other organisms such as bacteria and yeast (Ho et al. 2013). Bioethanol is produced from microalgal biomass by two processes: fermentation (biochemical process) and gasification (thermochemical process) (Ho et al. 2013). While conventional carbohydrate-rich crops (e.g. sugar beet, sugar cane, corn, cassava, lignocellulosic plants, etc.) must be cultivated on agricultural land to obtain optimum yield, microalgae can grow well on marginal land not suitable for agriculture. In addition, microalgal biomass has large quantity of starch and cellulose and lack lignin, a recalcitrant material that makes bioconversion challenging (Odjadjare et al. 2017). When compared to other chemical contents such as protein and lipids, carbohydrate levels in microalgal biomass are generally low, around 5–23% of dry weight. However, under stress conditions, certain strains of microalgae are known to accumulate carbohydrate in large amounts (Koutra et al. 2018; Vadiveloo et al. 2015). Many growth conditions also affect carbohydrate contents of the algae biomass (Ogbonna and Tanaka 1996).

The production of bioethanol from biomass of microalgae grown in wastewater is yet to be reported; nevertheless, impacts of various effluents on the carbohydrate content need to be investigated. Bio-based harvesting methods such as tannin-assisted coagulation-flocculation is shown to have a positive impact on the carbohydrate content relative to centrifugation technique, as cultivation systems showed an apparent effect on carbohydrate accumulation (Michelon et al. 2016). Besides using microalgal carbohydrate for bioethanol production, the biomass can be converted to recyclable and biodegradable plastics. Recently, the catalytic conversion of plant-based sugars (e.g. fructose) into furandicarboxylic acid, a chemical building block for polyethylene furanoate has been investigated as a sustainable alternative to petroleum-based packaging materials. Based on techno-economic analysis, biofuel production from microalgae is currently not feasible. However, coupling technologies with the biorefinery pathway, involving the valorization of the entire biomass, could lead to the feasibility of microalgal biofuel technology (Nwoba et al. 2020b).

7.6.4 Biopigments

Microalgal pigments primarily function to regulate the photosynthetic activity and impart colour to microalgae, and they include chlorophylls (a, b, c, d, f), carotenoids (astaxanthin, lutein, carotene, etc.), and phycobiliproteins (phycocyanin, allophycocyanin, phycoerythrin) (Nwoba et al. 2019b, 2020a, c). When compared with other components of the biomass such as carbohydrate, lipid, and protein, their productivity is much lower. Microalgal pigments are used in the production of colourants for food, fish and animal as well as supplements in food (Borowitzka 2013). The market value of the food-grade blue pigment, C-phycocyanin is US$500 per kg, while the reagent/analytical grade sells for US$125,000 per kg (Nwoba et al. 2019c). These pigments find applications in pharmaceutical, nutraceutical, and cosmetic industries, where they are exploited for their antibiotic, anti-tumour, anti-inflammatory, neuro-protective, anti-ageing, and functional food properties (Nwoba et al. 2019c). Some microalgae that are major players in pigments production are *Spirulina* sp. (phycocyanin, phycoerythrin), *Haematococcus pluvialis* (astaxanthin), *Dunaliella salina* (carotenoid), *Chlorella* sp. (chlorophylls), *Muriellopsis* sp. (lutein), and many others (Nwoba et al. 2020a). Microalgal pigment production using wastewaters can help to reduce the cost of nutrients input with the potential to increasing its productivity. Under limited light conditions as prevalent in wastewater due to its turbidity, enhanced production of pigments is usually observed in order to meet up with photosynthetic needs (Vadiveloo et al. 2015; Nwoba et al. 2016). Similarly, the concentration of pigments such as β-carotene can increase under limited nitrogen condition, while chlorophylls are known to be accumulated in high nitrogen content wastewaters (Michelon et al. 2016).

7.6.5 Proteins

Protein is an indispensable component of food, and its new sources are being sought for to be able to feed an additional 2.3 billion people on earth in 2050. Microalgae have emerged as a rich and novel protein source with capacity to address food crisis owing to the increasing human population. Utilizing wastewater to farm microalgae with the goal of harnessing the biomass as alternative feed additives for animals is a step towards sustainable agriculture (Batista et al. 2013; Van der Spiegel et al. 2013). Several microalgae species have high protein contents, over 46% of dry weight of biomass (e.g. *Chlorella* sp.) when cultured in wastewater (Nwoba et al. 2016; Moheimani et al. 2018). It has been shown that the contents of protein and the amino acid quality of wastewater-grown microalgae are similar to those of soya bean and egg (Moheimani et al. 2018). However, the use of microalgae grown in wastewater as novel protein source is hampered by serious challenges such as safety, public health, and hygiene. In addition, the quality of the protein product could also be questioned due to potential contaminants in wastewater such as the presence of potentially toxic metals, toxins, allergens, pathogens, and herbicides, which are capable of contaminating the biomass. Other factors such as biomass digestibility and the nucleic acids content constitute serious concerns. Hence, wastewater generated biomass can be utilized for animal feed if it meets regulatory permissible levels of the contaminants. For human application, it must be subjected to a stricter and thorough approval process before going to the market.

7.6.6 Polyhydroxyalkanoates (PHAs)

PHAs are natural polymers synthesized by heterotrophic bacteria (Troschl et al. 2017; Liu et al. 2016). They are highly biodegradable and biocompatible when compared to industrial plastics such as polyethylene and polypropylene. An example of this biopolymer is the polyhydroxybutyrate (PHB), which is biosynthesized intracellularly by microbial cells and the most commercial biopolymer (Troschl et al. 2017). Various microalgae in the cyanobacteria family such as *Spirulina platensis*, *Synechocystis* sp., *Nostoc muscorum*, etc. are well-known for their ability to produce PHB in large amount (29%–85% dry weight) in the biomass under mixotrophic culture conditions (Troschl et al. 2017; Koutra et al. 2018). Furthermore, PHB accumulation in some species can be induced by nutrient starvation especially nitrogen limitation and carbon replete conditions to reach 80% of biomass dry weight (Koutra et al. 2018). Using wastewater as a low-cost medium to cultivate cyanobacteria with the goal of producing PHB is a sustainable route towards bioplastic production. In this context, *Synechocystis* sp. has been cultivated in wastewater for PHB production, and results showed that the product yield was similar to those of bacteria, with stellar thermal and rheological qualities similar to commercial bioplastics (Kovalcik et al. 2017).

7.6.7 Exopolysaccharides

The exopolysaccharides represent a group of high-molecular-weight biopolymers biosynthesized by microorganisms during growth and excreted into the culture media. Generally, exopolysaccharides have found application in food as thickeners for textural and qualitative purpose (Liu et al. 2016). Their applications in medical, biotechnological, and industrial sectors are recognized. Exopolysaccharides produced by microalgae are known for antioxidant, antibacterial, antitumorigenic, and immunomodulatory activities (Delattre et al. 2016). Further, their sustainability and environmental friendliness features have made them to be used as bioflocculants, in wastewater/water purification and microalgal biomass dewatering (Liu et al. 2016). Microalgae such as *Chlorella vulgaris*, *Porphyridium cruentum*, *Lyngbya majuscula*, and *Micractinium* sp. have been studied for exopolysaccharides production (Liu et al. 2016; Wang et al. 2014). However, only a few researches (Wang et al. 2014) have investigated the synthesis of these substances by microalgae cultivated in wastewater.

7.6.8 Biofertilizers

The continued growth in human population requires higher output from agriculture and structures that produce food in order to meet increased food demand. Large quantities of fertilizers in the form N and P are needed to support intensive agriculture. The traditional chemical fertilizers are manufactured through the Haber process, and this is energy-intensive, not eco-friendly, and thus, unsustainable. Biomass obtained from microalgae cultivated in wastewater can serve as a biofertilizer for crop production because of significant amount of N and P. Biomass from microalgae has been shown to have antifungal, antibacterial, and antiviral activities and can be used as a biocontrol agent for pathogens (Kulik 1995; Prasanna et al. 2015). Algal biomass has been found to have a similar nutrient property with chemical fertilizers and above all releases nutrients slowly while remaining eco-friendly. Application of bio-based fertilizers in crop production has been associated with enhanced growth, germination, and flowering (Garcia-Gonzalez and Sommerfeld 2016).

7.7 CONCLUSIONS

The integration of microalgae cultivation with agricultural and industrial productions as well as municipal wastewater treatment plants for the purposes of wastewater treatment is clearly a sustainable biotechnology. Intensive microalgal cultivation in wastewaters represents an effective approach to the bioremediation of effluents, while producing valuable biomass that can be valorized to many important commodities, including biofuels and high-value compounds. The optimum system and mode of cultivation for efficient wastewater treatment depends on the species of microalgae, the nature of the wastewater (type of pollutants and their concentrations, as well as the turbidity of the wastewater), and most importantly, the desired level of treatment.

REFERENCES

Abdel-Raouf N, Al-Homaidan A, Ibraheem I (2012a) Microalgae and wastewater treatment. *Saudi Journal of Biological Sciences* 19 (3):257–275.

Abdel-Raouf N, Al-Homaidan AA, Ibraheem IB (2012b) Microalgae and wastewater treatment. *Saudi Journal of Biological Sciences* 19 (3):257–275.

Abou-Shanab RA, Ji M-K, Kim H-C, Paeng K-J, Jeon B-H (2013) Microalgal species growing on piggery wastewater as a valuable candidate for nutrient removal and biodiesel production. *Journal of Environmental Management* 115:257–264.

Ahluwalia SS, Goyal D (2007) Microbial and plant derived biomass for removal of heavy metals from wastewater. *Bioresource Technology* 98 (12):2243–2257.

Aizawa K, Miyachi S (1986) Carbonic anhydrase and CO2 concentrating mechanisms in microalgae and cyanobacteria. *FEMS Microbiology Letters* 39 (3):215–233. Doi: 10.1016/0378-1097(86)90447-7.

Ajjawi I, Verruto J, Aqui M, Soriaga LB, Coppersmith J, Kwok K, Peach L, Orchard E, Kalb R, Xu W (2017) Lipid production in *Nannochloropsis gaditana* is doubled by decreasing expression of a single transcriptional regulator. *Nature Biotechnology* 35 (7): 647–652.

Alhamed YA, Edris GM, GadelHak YM (2014) Performance evaluation of a bubble column photobioreactor for carbon dioxide sequestration by *Chlorella vulgaris*. *Arabian Journal for Science and Engineering* 39 (12):8453–8463.

Arias DM, Solé-Bundó M, Garfí M, Ferrer I, García J, Uggetti E (2018) Integrating microalgae tertiary treatment into activated sludge systems for energy and nutrients recovery from wastewater. *Bioresource Technology* 247:513–519.

Barsanti L, Gualtieri P (2014) *Algae: Anatomy, Biochemistry, and Biotechnology*. CRC Press, Boca Raton, FL.

Batista AP, Gouveia L, Bandarra NM, Franco JM, Raymundo A (2013) Comparison of microalgal biomass profiles as novel functional ingredient for food products. *Algal Research* 2 (2):164–173.

Becker EW (1994) *Microalgae, Biotechnology and Microbiology*. Cambridge University Press, Cambridge.

Benemann JR (1979) Production of nitrogen fertilizer with nitrogen-fixing blue - green algae. *Enzyme and Microbial Technology* 1 (2):83–90. Doi: 10.1016/0141-0229(79)90103-0.

Bhaya D, Schwarz R, Grossman R (2000) Molecular responses to environmental stress. In: Whitton BA, Potts M (eds) *The Ecology of Cyanobacteria Their Diversity in Time and Space*. Kluwer, Dordrecht, pp 397–442.

Borowitzka MA (1998) Limits to growth. In: Wong Y-S, Tam NFY (eds) *Wastewater Treatment with Algae*. Springer Berlin Heidelberg, Berlin, Heidelberg, pp 203–226. Doi: 10.1007/978-3-662-10863-5_12.

Borowitzka MA (2013) High-value products from microalgae—their development and commercialisation. *Journal of Applied Phycology* 25 (3):743–756.

Bouarab L, Dauta A, Loudiki M (2004) Heterotrophic and mixotrophic growth of *Micractinium pusillum* Fresenius in the presence of acetate and glucose: effect of light and acetate gradient concentration. *Water Research* 38 (11):2706–2712.

Brown MR, Dunstan GA, Norwood S, Miller KA (1996) Effects of harvest stage and light on the biochemical composition of the diatom *Thalassiosira pseudonana. Journal of Phycology* 32 (1):64–73.

Cai T, Ge X, Park SY, Li Y (2013a) Comparison of Synechocystis sp. PCC6803 and Nannochloropsis salina for lipid production using artificial seawater and nutrients from anaerobic digestion effluent. *Bioresource Technology* 144:255–260.

Cai T, Park SY, Li Y (2013b) Nutrient recovery from wastewater streams by microalgae: status and prospects. *Renewable and Sustainable Energy Reviews* 19:360–369. Doi: 10.1016/j.rser.2012.11.030.

Cai T, Park SY, Racharaks R, Li Y (2013c) Cultivation of Nannochloropsis salina using anaerobic digestion effluent as a nutrient source for biofuel production. *Applied Energy* 108:486–492.

Carpenter MJ (1987) *Biodegradation of an Oily Bilge Waste Using Algae.* Oklahoma Univ., Norman, OK.

Chen C-Y, Chang H-W, Kao P-C, Pan J-L, Chang J-S (2012) Biosorption of cadmium by CO_2-fixing microalga *Scenedesmus obliquus* CNW-N. *Bioresource Technology* 105:74–80.

Chevalier P, Proulx D, Lessard P, Vincent WF, de la Noüe J (2000) Nitrogen and phosphorus removal by high latitude mat-forming cyanobacteria for potential use in tertiary wastewater treatment. *Journal of Applied Phycology* 12 (2):105–112. Doi: 10.1023/A:1008168128654.

Chisti Y (2007) Biodiesel from microalgae. *Biotechnology Advances* 25 (3):294–306.

Chuka-ogwude D, Ogbonna J, Moheimani NR (2020) A review on microalgal culture to treat anaerobic digestate food waste effluent. *Algal Research* 47:101841.

Craggs R, Sutherland D, Campbell H (2012) Hectare-scale demonstration of high rate algal ponds for enhanced wastewater treatment and biofuel production. *Journal of Applied Phycology* 24 (3):329–337.

Delattre C, Pierre G, Laroche C, Michaud P (2016) Production, extraction and characterization of microalgal and cyanobacterial exopolysaccharides. *Biotechnology Advances* 34 (7):1159–1179.

de Wilt A, Butkovskyi A, Tuantet K, Leal LH, Fernandes TV, Langenhoff A, Zeeman G (2016) Micropollutant removal in an algal treatment system fed with source separated wastewater streams. *Journal of Hazardous Materials* 304:84–92. Doi: 10.1016/j.jhazmat.2015.10.033.

Dhodapkar RS, Gandhi KN (2019) Pharmaceuticals and personal care products in aquatic environment: chemicals of emerging concern? In: Prasad MNV, Vithanage M, Kapley A (eds) *Pharmaceuticals and Personal Care Products: Waste Management and Treatment Technology*. Butterworth-Heinemann Ltd, Oxford, pp 63–85.

Ding GT, Yaakob Z, Takriff MS, Salihon J, Abd Rahaman MS (2016) Biomass production and nutrients removal by a newly-isolated microalgal strain *Chlamydomonas sp* in palm oil mill effluent (POME). *International Journal of Hydrogen Energy* 41 (8):4888–4895.

Doria E, Longoni P, Scibilia L, Iazzi N, Cella R, Nielsen E (2012) Isolation and characterization of a *Scenedesmus acutus* strain to be used for bioremediation of urban wastewater. *Journal of Applied Phycology* 24 (3):375–383.

Garcia-Gonzalez J, Sommerfeld M (2016) Biofertilizer and biostimulant properties of the microalga *Acutodesmus dimorphus. Journal of Applied Phycology* 28 (2):1051–1061.

Gerba CP, Pepper IL (2019). Municipal wastewater treatment. In: Brusseau ML, Pepper IL, Gerba CP (eds) *Environmental and Pollution Science,* Academic Press, London, pp 393–418.

Gokulan R, Sathish N, Kumar RP (2013) Treatment of grey water using hydrocarbon producing *Botryococcus braunii. International Journal of ChemTech Research* 5 (3):1390–1392.

González LE, Cañizares RO, Baena S (1997) Efficiency of ammonia and phosphorus removal from a Colombian agroindustrial wastewater by the microalgae *Chlorella vulgaris* and *Scenedesmus dimorphus. Bioresource Technology* 60 (3):259–262.

Gopinath A, Puhan S, Nagarajan G (2009) Relating the cetane number of biodiesel fuels to their fatty acid composition: a critical study. *Proceedings of the Institution of Mechanical Engineers, Part D: Journal of Automobile Engineering* 223 (4):565–583.

Graham EJS, Dean CA, Yoshida TM, Twary SN, Teshima M, Alvarez MA, Zidenga T, Heikoop JM, Perkins GB, Rahn TA (2017) Oil and gas produced water as a growth medium for microalgae cultivation: a review and feasibility analysis. *Algal Research* 24:492–504.

Hawley A, Fallowfield H (2018) Pond walls: inclined planes to improve pathogen removal in pond systems for wastewater treatment? *Water Science and Technology* 78 (1):31–36.

Ho S-H, Huang S-W, Chen C-Y, Hasunuma T, Kondo A, Chang J-S (2013) Bioethanol production using carbohydrate-rich microalgae biomass as feedstock. *Bioresource Technology* 135:191–198.

Hongyang S, Yalei Z, Chunmin Z, Xuefei Z, Jinpeng L (2011) Cultivation of *Chlorella pyrenoidosa* in soybean processing wastewater. *Bioresource Technology* 102 (21):9884–9890.

Hu B, Min M, Zhou W, Li Y, Mohr M, Cheng Y, Lei H, Liu Y, Lin X, Chen P, Ruan R (2012) Influence of exogenous CO(2) on biomass and lipid accumulation of microalgae *Auxenochlorella protothecoides* cultivated in concentrated municipal wastewater. *Applied Biochemistry and Biotechnology* 166 (7):1661–1673. Doi: 10.1007/s12010-012-9566-2.

Hu Q, Sommerfeld M, Jarvis E, Ghirardi M, Posewitz M, Seibert M, Darzins A (2008) Microalgal triacylglycerols as feedstocks for biofuel production: perspectives and advances. *The plant Journal* 54 (4):621–639.

Huerlimann R, De Nys R, Heimann K (2010) Growth, lipid content, productivity, and fatty acid composition of tropical microalgae for scale-up production. *Biotechnology and Bioengineering* 107 (2):245–257.

Idenyi JN, Ubi BE, Nwoba EG, Ebenyi N, Oko AO, Okah YO (2019) Efficiency of local *Chlorella* species in wastewater treatment. Foundation for African Development through International Biotechnology (FADIB). *Proceedings of the 25th annual international conference/workshop Biotechnology: Catalyst for Sustainable Development*, Enugu, Nigeria.

Ishika T, Moheimani NR, Bahri PA (2017) Sustainable saline microalgae co-cultivation for biofuel production: a critical review. *Renewable and Sustainable Energy Reviews* 78:356–368.

Islam MA, Magnusson M, Brown RJ, Ayoko GA, Nabi MN, Heimann K (2013) Microalgal species selection for biodiesel production based on fuel properties derived from fatty acid profiles. *Energies* 6 (11):5676–5702.

Jaysudha S, Sampathkumar P (2014) Nutrient removal from tannery effluent by free and immobilized cells of marine microalgae *Chlorella salina. International Journal of Environmental Biology* 4 (1):21–26.

Jimenez-Perez M, Sanchez-Castillo P, Romera O, Fernandez-Moreno D, Pérez-Martınez C (2004) Growth and nutrient removal in free and immobilized planktonic green algae isolated from pig manure. *Enzyme and Microbial Technology* 34 (5):392–398.

Kamyab H, Chelliapan S, Din MFM, Shahbazian-Yassar R, Rezania S, Khademi T, Kumar A, Azimi M (2017) Evaluation of *Lemna minor* and *Chlamydomonas* to treat palm oil mill effluent and fertilizer production. *Journal of Water Process Engineering* 17:229–236.

Kesaano M, Sims RC (2014) Algal biofilm based technology for wastewater treatment. *Algal Research* 5:231–240.

Khona DK, Shirolikar SM, Gawde KK, Hom E, Deodhar MA, D'Souza JS (2016) Characterization of salt stress-induced palmelloids in the green alga, *Chlamydomonas reinhardtii*. *Algal Research* 16:434–448. Doi: 10.1016/j.algal.2016.03.035.

Kim G-Y, Yun Y-M, Shin H-S, Kim H-S, Han J-I (2015) *Scenedesmus*-based treatment of nitrogen and phosphorus from effluent of anaerobic digester and bio-oil production. *Bioresource Technology* 196:235–240.

Kimura K, Hane Y, Watanabe Y, Amy G, Ohkuma N (2004) Irreversible membrane fouling during ultrafiltration of surface water. Water Research 38(14–15):3431–3441.

Knothe G (2009) Improving biodiesel fuel properties by modifying fatty ester composition. *Energy & Environmental Science* 2 (7):759–766.

Kong W, Song H, Cao Y, Yang H, Hua S, Xia C (2011) The characteristics of biomass production, lipid accumulation and chlorophyll biosynthesis of *Chlorella vulgaris* under mixotrophic cultivation. *African Journal of Biotechnology* 10 (55):11620–11630.

Koutra E, Economou CN, Tsafrakidou P, Kornaros M (2018) Bio-based products from microalgae cultivated in digestates. *Trends in Biotechnology* 36 (8):819–833.

Kovalcik A, Meixner K, Mihalic M, Zeilinger W, Fritz I, Fuchs W, Kucharczyk P, Stelzer F, Drosg B (2017) Characterization of polyhydroxyalkanoates produced by *Synechocystis salina* from digestate supernatant. *International Journal of Biological Macromolecules* 102:497–504.

Kulik MM (1995) The potential for using cyanobacteria (blue-green algae) and algae in the biological control of plant pathogenic bacteria and fungi. *European Journal of Plant Pathology* 101 (6):585–599.

Kwietniewska E, Tys J (2014) Process characteristics, inhibition factors and methane yields of anaerobic digestion process, with particular focus on microalgal biomass fermentation. *Renewable and Sustainable Energy Reviews* 34:491–500.

Larsdotter K (2006) Wastewater treatment with microalgae-a literature review. *Vatten* 62 (1):31.

Li K, Liu Q, Fang F, Luo R, Lu Q, Zhou W, Huo S, Cheng P, Liu J, Addy M, Chen P, Chen D, Ruan R (2019) Microalgae-based wastewater treatment for nutrients recovery: a review. *Bioresource Technology* 291:121934. Doi: 10.1016/j.biortech.2019.121934.

Li Y, Chen Y-F, Chen P, Min M, Zhou W, Martinez B, Zhu J, Ruan R (2011a) Characterization of a microalga *Chlorella* sp. well adapted to highly concentrated municipal wastewater for nutrient removal and biodiesel production. *Bioresource Technology* 102 (8):5138–5144.

Li Y, Zhou W, Hu B, Min M, Chen P, Ruan RR (2011b) Integration of algae cultivation as biodiesel production feedstock with municipal wastewater treatment: strains screening and significance evaluation of environmental factors. *Bioresource Technology* 102 (23):10861–10867. Doi: 10.1016/j.biortech.2011.09.064.

Li Y, Zhou W, Hu B, Min M, Chen P, Ruan RR (2012) Effect of light intensity on algal biomass accumulation and biodiesel production for mixotrophic strains *Chlorella kessleri* and *Chlorella protothecoide* cultivated in highly concentrated municipal wastewater. *Biotechnology and Bioengineering* 109 (9):2222–2229. Doi: 10.1002/bit.24491.

Liang Z, Liu Y, Ge F, Xu Y, Tao N, Peng F, Wong M (2013) Efficiency assessment and pH effect in removing nitrogen and phosphorus by algae-bacteria combined system of *Chlorella vulgaris* and *Bacillus licheniformis*. *Chemosphere* 92 (10):1383–1389. Doi: 10.1016/j.chemosphere.2013.05.014.

Lim S-L, Chu W-L, Phang S-M (2010) Use of *Chlorella vulgaris* for bioremediation of textile wastewater. *Bioresource Technology* 101 (19):7314–7322.

Liu L, Pohnert G, Wei D (2016) Extracellular metabolites from industrial microalgae and their biotechnological potential. *Marine Drugs* 14 (10):191.

Liu X, Duan S, Li A, Xu N, Cai Z, Hu Z (2009) Effects of organic carbon sources on growth, photosynthesis, and respiration of *Phaeodactylum tricornutum*. *Journal of Applied Phycology* 21 (2):239–246.

Lundquist TJ, Woertz IC, Quinn N, Benemann JR (2010) A realistic technology and engineering assessment of algae biofuel production. *Energy Biosciences Institute* 1:1–178.

Mehrabadi A, Craggs R, Farid MM (2015) Wastewater treatment high rate algal ponds (WWT HRAP) for low-cost biofuel production. *Bioresource Technology* 184:202–214. Doi: 10.1016/j.biortech.2014.11.004.

Miao X, Wu Q (2006) Biodiesel production from heterotrophic microalgal oil. *Bioresource Technology* 97 (6):841–846. Doi: 10.1016/j.biortech.2005.04.008.

Michelon W, Da Silva MLB, Mezzari MP, Pirolli M, Prandini JM, Soares HM (2016) Effects of nitrogen and phosphorus on biochemical composition of microalgae polyculture harvested from phycoremediation of piggery wastewater digestate. *Applied Biochemistry and Biotechnology* 178 (7):1407–1419.

Milano J, Ong HC, Masjuki H, Chong W, Lam MK, Loh PK, Vellayan V (2016) Microalgae biofuels as an alternative to fossil fuel for power generation. *Renewable and Sustainable Energy Reviews* 58:180–197.

Min M, Hu B, Zhou W, Li Y, Chen P, Ruan R (2012) Mutual influence of light and CO_2 on carbon sequestration via cultivating mixotrophic alga *Auxenochlorella protothecoides* UMN280 in an organic carbon-rich wastewater. *Journal of Applied Phycology* 24 (5):1099–1105. Doi: 10.1007/s10811-011-9739-3.

Miranda AF, Ramkumar N, Andriotis C, Höltkemeier T, Yasmin A, Rochfort S, Wlodkowic D, Morrison P, Roddick F, Spangenberg G (2017) Applications of microalgal biofilms for wastewater treatment and bioenergy production. *Biotechnology for Biofuels* 10 (1):120.

Moheimani NR, Vadiveloo A, Ayre JM, Pluske JR (2018) Nutritional profile and in vitro digestibility of microalgae grown in anaerobically digested piggery effluent. *Algal Research* 35:362–369.

Moser BR (2014) Impact of fatty ester composition on low temperature properties of biodiesel–petroleum diesel blends. *Fuel* 115:500–506.

Muttamara S (1996) Wastewater characteristics. *Resources, Conservation and Recycling* 16 (1–4): 145–159.

Neff J, Sauer T, Hart A (2011) Bioaccumulation of hydrocarbons from produced water discharged to offshore waters of the US Gulf of Mexico. In: Lee K, Neff JM (eds) *Produced Water: Environmental Risks and Advances in Mitigation Technologies*. Springer, New York, pp 441–477.

Neveux N, Bolton JJ, Bruhn A, Roberts DA, Ras M (2018) The bioremediation potential of seaweeds: recycling nitrogen, phosphorus, and other waste products. *Blue Biotechnology: Production and Use of Marine Molecules* 1: 217–241.

Nwoba EG, Ayre JM, Moheimani NR, Ubi BE, Ogbonna JC (2016) Growth comparison of microalgae in tubular photobioreactor and open pond for treating anaerobic digestion piggery effluent. *Algal Research* 17:268–276. Doi: 10.1016/j.algal.2016.05.022.

Nwoba EG, Mickan BS, Moheimani NR (2019a) *Chlorella* sp. growth under batch and fed-batch conditions with effluent recycling when treating the effluent of food waste anaerobic digestate. *Journal of Applied Phycology* 31 (6):3545–3556.

Nwoba EG, Ogbonna CN, Ishika T, Vadiveloo A (2020a) Microalgal pigments: a source of natural food colors. In: Alam MA, Xu J-L, Wang Z (eds) *Microalgae Biotechnology for Food, Health and High Value Products*. Springer, Singapore, pp 81–123.

Nwoba EG, Parlevliet DA, Laird DW, Alameh K, Louveau J, Pruvost J, Moheimani NR (2020b) Energy efficiency analysis of outdoor standalone photovoltaic-powered photobioreactors coproducing lipid-rich algal biomass and electricity. *Applied Energy* 275:115403.

Nwoba EG, Parlevliet DA, Laird DW, Alameh K, Moheimani NR (2019b) Light management technologies for increasing algal photobioreactor efficiency. *Algal Research* 39:101433.

Nwoba EG, Parlevliet DA, Laird DW, Alameh K, Moheimani NR (2019c) Sustainable phycocyanin production from *Arthrospira platensis* using solar-control thin film coated photobioreactor. *Biochemical Engineering Journal* 141:232–238.

Nwoba EG, Parlevliet DA, Laird DW, Alameh K, Moheimani NR (2020c) Outdoor phycocyanin production in a standalone thermally-insulated photobioreactor. *Bioresource Technology* 315:123865.

Nwoba EG, Parlevliet DA, Laird DW, Alameh K, Moheimani NR (2020d) Pilot-scale self-cooling microalgal closed photobioreactor for biomass production and electricity generation. *Algal Research* 45:101731.

Nwoba EG, Vadiveloo A, Ogbonna CN, Ubi BE, Ogbonna JC, Moheimani NR (2020e) Algal cultivation for treating wastewater in African developing countries. A review. *CLEAN–Soil, Air, Water* 48(3):2000052.

Odjadjare EC, Mutanda T, Olaniran AO (2017) Potential biotechnological application of microalgae: a critical review. *Critical Reviews in Biotechnology* 37 (1):37–52.

Ogbonna J, Ichige E, Tanaka H (2002a) Interactions between photoautotrophic and heterotrophic metabolism in photoheterotrophic cultures of *Euglena gracilis. Applied Microbiology and Biotechnology* 58 (4):532–538.

Ogbonna JC, Ichige E, Tanaka H (2002b) Regulating the ratio of photoautotrophic to heterotrophic metabolic activities in photoheterotrophic culture of *Euglena gracilis* and its application to α-tocopherol production. *Biotechnology Letters* 24 (12):953–958.

Ogbonna JC, Masui H, Tanaka H (1997) Sequential heterotrophic/autotrophic cultivation–an efficient method of producing *Chlorella* biomass for health food and animal feed. *Journal of Applied Phycology* 9 (4):359–366.

Ogbonna JC, McHenry MP (2015) Culture systems incorporating heterotrophic metabolism for biodiesel oil production by microalgae. In: Moheimani NR, McHenry MP, de Boer k, Bahri PA (eds) *Biomass and Biofuels from Microalgae* . Springer, Cham, pp 63–74.

Ogbonna JC, Moheimani NR (2015) Potentials of exploiting heterotrophic metabolism for biodiesel oil production by microalgae. In: Moheimani NR, McHenry MP, de Boer k, Bahri PA (eds) *Biomass and Biofuels from Microalgae.* Springer, Cham, pp 45–61.

Ogbonna JC, Tomiyama S, Tanaka H (1999) Production of α-tocopherol by sequential heterotrophic-photoautotrophic cultivation of *Euglena gracilis.* In: Osinga R, Tramper J, Burgess JG, Wijffels RH (eds) *Progress in Industrial Microbiology*, vol 35. Elsevier, Noordwijkerhout, pp 213–221.

Ogbonna JC, Tomiyamal S, Tanaka H (1998) Heterotrophic cultivation of *Euglena gracilis* Z for efficient production of α-tocopherol. *Journal of Applied Phycology* 10 (1):67–74.

Ogbonna JC, Yoshizawa H, Tanaka H (2000) Treatment of high strength organic wastewater by a mixed culture of photosynthetic microorganisms. *Journal of Applied Phycology* 12 (3–5):277–284.

Olguín EJ (2012) Dual purpose microalgae–bacteria-based systems that treat wastewater and produce biodiesel and chemical products within a Biorefinery. *Biotechnology Advances* 30 (5):1031–1046. Doi: 10.1016/j.biotechadv.2012.05.001.

Oliver RL, Ganf GG (2002) Freshwater blooms. In:. Whitton BA, Potts M (eds) *The Ecology of Cyanobacteria: Their Diversity in Time and Space.* Springer Science and Business media, Cham.

Park JB, Craggs RJ (2010) Wastewater treatment and algal production in high rate algal ponds with carbon dioxide addition. *Water Science and Technology* 61 (3):633–639. Doi: 10.2166/wst.2010.951.

Park JB, Craggs RJ, Shilton AN (2011) Wastewater treatment high rate algal ponds for biofuel production. *Bioresource Technology* 102 (1):35–42. Doi: 10.1016/j.biortech.2010.06.158.

Peng FQ, Ying GG, Yang B, Liu S, Lai HJ, Liu YS, Chen ZF, Zhou GJ (2014) Biotransformation of progesterone and norgestrel by two freshwater microalgae (Scenedesmus obliquus and Chlorella pyrenoidosa): transformation kinetics and products identification. *Chemosphere* 95:581–588. Doi: 10.1016/j.chemosphere.2013.10.013.

Prajapati SK, Kumar P, Malik A, Vijay VK (2014) Bioconversion of algae to methane and subsequent utilization of digestate for algae cultivation: a closed loop bioenergy generation process. *Bioresource Technology* 158:174–180.

Prasanna R, Babu S, Bidyarani N, Kumar A, Triveni S, Monga D, Mukherjee AK, Kranthi S, Gokte-Narkhedkar N, Adak A (2015) Prospecting cyanobacteria-fortified composts as plant growth promoting and biocontrol agents in cotton. *Experimental Agriculture* 51 (1):42–65.

Rajkumara R, Takriffab M (2015) Nutrient removal from anaerobically treated palm oil mill effluent by *Spirulina platensis* and *Scenedesmus dimorphus*. *Der Pharmacia Lettre* 7 (7):416–421.

Rao P, Kumar RR, Raghavan B, Subramanian V, Sivasubramanian V (2011) Application of phycoremediation technology in the treatment of wastewater from a leather-processing chemical manufacturing facility. *Water Sa* 37 (1):7–14.

Rayen F, Behnam T, Dominique P (2019) Optimization of a raceway pond system for wastewater treatment: a review. *Critical Reviews in Biotechnology* 39 (3):422–435.

Richmond A (2008) *Handbook of Microalgal Culture: Biotechnology and Applied Phycology*. John Wiley & Sons, New York.

Richmond A, Hu Q (2013) *Handbook of Microalgal Culture: Applied Phycology and Biotechnology*. John Wiley & Sons, Chichester.

Ruiz-Martínez A, Serralta J, Romero I, Seco A, Ferrer J (2015) Effect of intracellular P content on phosphate removal in *Scenedesmus* sp. Experimental study and kinetic expression. *Bioresource Technology* 175:325–332. Doi: 10.1016/j.biortech.2014.10.081.

Sawayama S, Rao K, Hall D (1998) Nitrate and phosphate ion removal from water by *Phormidium laminosum* immobilized on hollow fibres in a photobioreactor. *Applied Microbiology and Biotechnology* 49 (4):463–468.

Schenk PM, Thomas-Hall SR, Stephens E, Marx UC, Mussgnug JH, Posten C, Kruse O, Hankamer B (2008) Second generation biofuels: high-efficiency microalgae for biodiesel production. *Bioenergy Research* 1 (1):20–43.

Sharma GK, Khan SA (2013) Bioremediation of sewage wastewater using selective algae for manure production. *International Journal of Environmental Engineering and Management* 4 (6):573–580.

Shin DY, Cho HU, Utomo JC, Choi Y-N, Xu X, Park JM (2015) Biodiesel production from *Scenedesmus bijuga* grown in anaerobically digested food wastewater effluent. *Bioresource Technology* 184:215–221.

Shuba ES, Kifle D (2018) Microalgae to biofuels: 'promising' alternative and renewable energy, review. *Renewable and Sustainable Energy Reviews* 81:743–755.

Sims RE, Mabee W, Saddler JN, Taylor M (2010) An overview of second generation biofuel technologies. *Bioresource Technology* 101 (6):1570–1580.

Smith VH (2003) Eutrophication of freshwater and coastal marine ecosystems a global problem. *Environmental Science and Pollution Research* 10 (2):126–139.

Soetrisnanto D, Hadiyanto H (2014) Phytoremediation of palm oil mill effluent by using *Pistia stratiotes* plant and *Algae Spirulina* sp for biomass production (RESEARCH NOTE). *International Journal of Engineering* 27 (12):1809–1814.

Sun Z-l, Sun L-q, Chen G-z (2019) Microalgal cultivation and nutrient removal from digested piggery wastewater in a thin-film flat plate photobioreactor. *Applied Biochemistry and Biotechnology* 187 (4):1488–1501.

Tamil Selvam TB, Renganathan R, Takriff MS (2015) Nutrient removal of POME using POME isolated microalgae strain, *Characium* sp. In: *Advanced Materials Research*, Trans Tech Publications Ltd, Freienbach, pp 364–369.

Troschl C, Meixner K, Drosg B (2017) Cyanobacterial PHA production—Review of recent advances and a summary of three years' working experience running a pilot plant. *Bioengineering* 4 (2):26.

Udom I, Zaribaf BH, Halfhide T, Gillie B, Dalrymple O, Zhang Q, Ergas SJ (2013) Harvesting microalgae grown on wastewater. *Bioresource Technology* 139:101–106.

Vadiveloo A, Moheimani NR, Cosgrove JJ, Bahri PA, Parlevliet D (2015) Effect of different light spectra on the growth and productivity of acclimated *Nannochloropsis* sp. (Eustigmatophyceae). *Algal Research* 8:121–127.

Vadiveloo A, Nwoba EG, Ogbonna C, Mehta P (2019) Sustainable production of bioproducts from wastewater-grown microalgae. In Sustainable Downstream Processing of Microalgae for Industrial Application. CRC Press, Boca Raton, pp 165–200

Van der Spiegel M, Noordam M, Van der Fels-Klerx H (2013) Safety of novel protein sources (insects, microalgae, seaweed, duckweed, and rapeseed) and legislative aspects for their application in food and feed production. *Comprehensive Reviews in Food Science and Food Safety* 12 (6):662–678.

Vidali M (2001) Bioremediation. An overview. *Pure and Applied Chemistry* 73 (7):1163–1172.

Wang H, Zhang W, Chen L, Wang J, Liu T (2013) The contamination and control of biological pollutants in mass cultivation of microalgae. *Bioresource Technology* 128:745–750.

Wang L, Min M, Li Y, Chen P, Chen Y, Liu Y, Wang Y, Ruan R (2010) Cultivation of green algae *Chlorella* sp. in different wastewaters from municipal wastewater treatment plant. *Applied Biochemistry Biotechnology* 162 (4):1174–1186. Doi:10.1007/s12010-009-8866-7.

Wang M, Kuo-Dahab WC, Dolan S, Park C (2014) Kinetics of nutrient removal and expression of extracellular polymeric substances of the microalgae, *Chlorella* sp. and *Micractinium* sp., in wastewater treatment. *Bioresource Technology* 154:131–137.

Ward A, Lewis D, Green F (2014) Anaerobic digestion of algae biomass: a review. *Algal Research* 5:204–214.

Xiong J-Q, Kurade MB, Jeon B-H (2018) Can Microalgae Remove Pharmaceutical Contaminants from Water? *Trends in Biotechnology* 36 (1):30–44. Doi: 10.1016/j.tibtech.2017.09.003.

Xu Z, Yan X, Pei L, Luo Q, Xu J (2008) Changes in fatty acids and sterols during batch growth of *Pavlova viridis* in photobioreactor. *Journal of Applied Phycology* 20 (3):237–243.

Yadavalli R, Heggers GRVN (2013) Two stage treatment of dairy effluent using immobilized *Chlorella pyrenoidosa*. *Journal of Environmental Health Science and Engineering* 11 (1):36.

Yu Z, Song M, Pei H, Han F, Jiang L, Hou Q (2017) The growth characteristics and biodiesel production of ten algae strains cultivated in anaerobically digested effluent from kitchen waste. *Algal Research* 24:265–275.

Zamalloa C, Vulsteke E, Albrecht J, Verstraete W (2011) The techno-economic potential of renewable energy through the anaerobic digestion of microalgae. *Bioresource Technology* 102 (2):1149–1158.

Zeng F, Huang J, Meng C, Zhu F, Chen J, Li Y (2016) Investigation on novel raceway pond with inclined paddle wheels through simulation and microalgae culture experiments. *Bioprocess and Biosystems Engineering* 39 (1):169–180.

Zheng H, Liu M, Lu Q, Wu X, Ma Y, Cheng Y, Addy M, Liu Y, Ruan R (2018) Balancing carbon/nitrogen ratio to improve nutrients removal and algal biomass production in piggery and brewery wastewaters. *Bioresource Technology* 249:479–486. Doi: 10.1016/j.biortech.2017.10.057.

Zhu L, Wang Z, Shu Q, Takala J, Hiltunen E, Feng P, Yuan Z (2013) Nutrient removal and biodiesel production by integration of freshwater algae cultivation with piggery wastewater treatment. *Water Research* 47 (13):4294–4302.

8 New Insights into Chlorine Disinfectant Usage in Water Quality

A Need for Review of Existing Guidelines

Mojisola Owoseni
Federal University of Lafia

Anthony I. Okoh
University of Sharjah
University of Fort Hare

CONTENTS

8.1 INTRODUCTION

Water plays a key role in all facets of life, in socio-economic development, food security and environmental development. It is also vital in the spiritual and aesthetic lives of billions of people (Martino, 2003). However, population explosion, hospital effluents and industrial discharges have contributed immensely to huge generation of sewage leading to pollution of waterbodies and scarcity of potable water (Green

DOI: 10.1201/9781003180289-8

et al., 2015). Existing effluent strategies for sewage treatment are grossly inadequate or non-existent, particularly in Sub-Saharan Africa, resulting in poorly treated effluent discharged to waterbodies thus reducing access to clean water and contributing to socio-economic and health deprivation of the populace (United Nations, 2015).

Earlier forms of water treatment included filtration using charcoal, exposure to sunlight, boiling and straining (Baker and Taras, 1981), the use of chemical alums by Egyptians in 1500 B.C. to precipitate out suspended particles. In the early 1900s, focus shifted to the use of disinfectants such as chlorine to improve water quality and to reduce the incidence of waterborne diseases (HDR, 2001).

The first documentation of water chlorination was carried out by John Snow in 1854 in London following an outbreak of cholera (White, 1999). Few years later, chlorine was used in the treatment for domestic water following typhoid outbreak in England by Sim Woodhead (Leal, 1909). The United States recorded her first usage of chlorine disinfectant for treatment of community water in New Jersey in 1908 with substantial reduction in the incidence of waterborne diseases including cholera, dysentery and typhoid over a century (CDC, 1999).

Statistical evidence from chlorine usage in the United States showed a decline in mortality rate from 36 per 100,000 to 20 per 100,000 between 1900 and 1910, 3 per 100,000 in 1935 with only 20 deaths from typhoid in 1960 and complete eradication of typhoid in the United States and other developed nations (Laubusch, 1964; Tiernan, 1948). Nevertheless, with over 40 years of continuous improvement in water treatment technologies, drinking water is still plagued with the onslaught of microbial pathogens in the United States thus increasing cost expenditure of treating water-related diseases of up to one billion annually (Yodder et al., 2008).

The overarching goal of water treatment is the production of quality water (depending on the end use) that requires minimum additional treatments to manage public health and environmental hazards (EPA, 2002). In order to curb the upsurge of pathogen survival in disinfected water, water treatment processes are regulated within the prescribed limits. However, over many decades, pathogens have evolved chlorine resistant processes to undermine the efficacy of chlorination (Martinez-Hernandez et al., 2013; Igbinosa et al., 2011; Okoh et al., 2010) which posits a serious hazard to human health and the environment.

High chlorine concentrations have been reported to contribute to antibiotic resistance and as precursors of disinfection-by-products (DBPs) when organic matter is present in high concentrations, affecting reproductive and developmental health and bladder cancer. Of particular public health concerns are the DBPs which are assumed to be mutagenic and genotoxic drivers of cancers by various epidemiological studies due to their carcinogenic nature. However, these studies suggest inadequate evidence to correlate high doses of chlorine and damage to reproductive and developmental health and onset of cancer development (Du et al., 2017). This review provides an overview of chlorine disinfectant usage in water quality at the recommended chlorine dosages and integrates the evidence of emerging chlorine resistant pathogens. This will help address the need for a review of current guidelines for optimization of chlorine disinfection in water treatment in balancing the acute risks involved in pathogen survival and the chronic risks of DBPs.

8.2 WATER TREATMENT PROCESSES

Water treatment processes follow four main stages of preliminary treatment, primary treatment, secondary treatment and tertiary treatment.

The preliminary treatment involves mechanical removal of large solid materials from influent wastewaters through the use of screen, the primary treatment targets the removal of settleable organic and inorganic solids by mechanical process through filtration and sedimentation processes (Sonune and Ghate, 2004) and the secondary treatment is a biological treatment stage that utilizes natural self-purification process to remove 90% of suspended material and biochemical oxygen demand (BOD).

Tertiary stage or disinfection of water is aimed at achieving the following purposes: to eliminate organisms at a required and desirable level and to maintain the maximum contaminant level (MCL) for DBPs set by Environmental Protection agency (EPA, 2002). Disinfectants are usually in physical or chemical forms utilized singly or in consortium with other disinfectants. Physical disinfectants include ultraviolet (UV) light, while chemical disinfectants include peracetic acid (PAA), hydrogen peroxide, ozone, pure chlorine, chlorine gas, chlorine dioxide and chlorine compounds such as Sodium hypochlorite and Calcium hypochlorite (Al-Adhami et al., 2007). UV radiation has been effectively used against viruses (Koivunen and Heinonen-Tanski, 2005). It penetrates the cell wall of microbial pathogens, forms thymine dimers with the DNA and inhibits cell replication and infectivity. However, at low doses, UV can lead to DNA reversal of pathogens through cellular repair mechanism such as photoreactivation, nucleotide excision repair or recombination repair. UV lamp is also fragile and contains toxic mercury that is hazardous to the environment (Chevremont et al., 2013), and no residue is formed after the disinfection process (EPA, 2011). Ozone alters cell wall permeability of bacterial pathogens and causes inactivation of DNA (Cho et al., 2010). Conversely, it is an expensive and energy consuming process (Lazarova et al., 1999). Ozone forms DBP carcinogen bromate ion in water containing bromide ions (Shannon et al., 2007). Chlorine is used in the treatment of water, as wash water disinfectant in food processing to minimize microbial density in fresh food, to improve shell life and surface disinfection in hospitals (Stopforth et al., 2008; Helbling et al; 2007), as point-of-use disinfectant in areas without access to municipally distributed water to improve water quality and to substantially reduce the risk of diarrhoea by 29% in children (Arnold and Colford Jr, 2007; Crump et al., 2005). The sensitivities of microorganisms to chlorine have been demonstrated in various studies. Chlorine concentration of 1.1 mg/L for 1 minute inactivated pathogenic and wild strains *E. coli* O157:H7 at approximately 4 log, while *Enterococcus*, *Enterobacter*, *Pseudomonas*, *Staphylococcus* and *Bacillus* were eliminated in chlorinated water (Bergeron et al., 2015).

On the other hand, chlorine is less effective against bacteria spores, viruses and protozoans including *Cryptosporidium* and *Giardia* (Bouki et al., 2013) and forms DBPs including trihalomethanes (THMs) and haloacetic acids (HAAs) on reaction with organic particles in water and the production of off-tastes and odours (Michael et al., 2013).

8.3 RECOMMENDED GUIDELINES FOR CHLORINE DISINFECTION IN WATER TREATMENT

In order to meet global and domestic needs, various guidelines have been set to standardize the chlorine disinfection usage. The WHO stipulates a minimum free chlorine concentration of 0.5 mg/L after 30 minutes at a pH of less than 8, and a turbidity of less than 1 Nephelometric Turbidity Unit (NTU) for drinking water and a residual free chlorine concentration between 0.5 and 1.0 mg/L in water at the point of discharge from the treatment plant (WHO, 2011; EPA, 2011). The performance of disinfection systems that form residual chlorine in water is commonly measured by the product of their concentration and time ($C \times t$) values and a *Ct* value of 15 mg/min/L is recommended for disinfection of portable water. This serves as a general prescribed limit whereas a more site-specific approach is recommended to setting *Ct* values (WHO, 2006).

Several countries have prescribed regulations for monitoring production of portable water and effluent discharge. For example, in Nigeria, the chlorine residual for drinking water ranges between 0.2 and 0.25 mg/L and 0 CFU/mL of *E. coli* and 10 CFU/mL of total coliform counts (NSDWQ, 2007), while no guideline exists for wastewater/sewage treatment. In South Africa, permissible discharge limits for free residual chlorine concentration in drinking water is expected to contain 0 CFU/100 mL of bacteria, while for wastewater effluents, 0.25 mg/L and 0 mg/L at special limits are specified with permissible limit of 1,000 CFU/100 mL of faecal coliform count (DWAF, 2013; WHO, 2011). The European water guidelines specified a parametric value of zero *E. coli*/100 mL in portable water (Council directive, 1998). The efficiency of water chlorination from empirical studies is compared to prescribed guidelines (Table 8.1). Chlorination is expected to cover a wide range of microbial groups. In order to achieve a reduction efficiency of Ct_{99} for viruses, it is expected that drinking water should be treated at 2–30 mg/min/L of chlorine at 0°C–10°C and at pH 7–9, for bacteria at 0.04–0.08 mg/min/L at 5°C and pH 6–7 and Protozoa (Giardia) at 25–245 mg/min/L at 0°C–25°C and pH 7–8 (WHO, 2011).

8.4 MECHANISMS OF ACTION OF CHLORINE IN WATER TREATMENT

The disinfectant-pathogen inactivation is the ability of the disinfectant to destroy the cellular structure of the microorganisms or disrupt its metabolic activities, biosynthesis, growth or reproduction. Disinfectants inactivate bacteria by preventing them from dividing and forming colonies, viruses from forming plaques in host cells and protozoan oocysts such as *Cryptosporidium* by inability to multiply and infect their hosts (EPA, 2011).

In aqueous environments, chlorine dissociates into unionized hypochlorous acid (HOCl), the more reactive form and hypochlorite ions (ClO^-), a reaction dependent on pH and temperature (Deborde and Gunten, 2008). HOCl rapidly induces the unfolding of protein in vitro and irreversible protein aggregation in vivo and in vitro which contributes to its highly bactericidal effect (Winter et al., 2008). HOCl damages the inner membrane and DNA, causes oxidative unfolding of the thiol group and protein aggregation at the cysteine and methionine residues leading to irreversible aggregation (Drazic et al., 2014; Leichert et al., 2007). It reacts with nucleotides, inactivates

Type of Water	Microorganism	VC/ Survival Rates	FC (mg/L)	Acceptable Limits			References
				VC/ Survival Rates	FC (mg/L)	CT (min)	
Drinking water	Rotavirus	1.44×10^2–2.23×10^4 PFU/L	0.65–0.84 mg/L	-	*0.3–4*[1] GB (2006)	> 30 CDC (2012)	Ye et al. (2012)
Wastewater	Faecal coliform/ *E. coli*	0.7–4.95 $\log_{10}$ CFU/mL; 1.3×10^2–2.8×10^3 CFU/100 mL	0.09–0.46; 0.05–0.71	0 CFU/mL[1], 0 CFU/100mL (relaxed limit)[2], <1,000 CFU/mL (general limit) (WHO, 2006)	*0.25*[2] DWAF (2013)	30 WHO (2006)	Dungeni et al. (2010); Osuolale and Okoh (2015)
Wastewater	Vibrio	(4.5×10^3–1.9×10^5; 2.1×10^1–1.8×10^5) CFU/mL	0.09–1.4; 0.07–3.85	0.5–2.0 (Vibrio cholera)	0.25[2] CDC (2012), DWAF (2013)	30 DWAF, (2013)	Igbinosa et al. (2011), Igbinosa and Okoh (2009)

CT, contact time; FC, residual chlorine; VC, viable count.

[1] portable water

[2] wastewater.

enzymes and inhibits cell transport (Finnegan et al., 2010). HOCl removes electrons from reacting chemical groups, oxidizes them and become reduced after the reaction (Rosen et al., 2009).

Several hypotheses surround the mechanisms of microbial inactivation by chlorine disinfectant. Cho et al. (2010) proposed that the mechanism of disinfection is characterized by two pathways. The primary disinfection step involves the alteration of the bacterial cell wall and spore coat integrity causing a physicochemical change in the cell surface. This is followed by a secondary damage to the intracellular constituents and functions, loss of proteins and nucleic acids including the degradation of approximately 76% of β-galactosidase enzyme at 1.0 log inactivation without significant alteration of the cell membrane and lipid perioxidation. The oxidation of the outer membrane and periplasmic proteins correlates poorly with chlorine inactivation (Drazic et al., 2014). Therefore, disruption of the cellular membrane is not regarded as a major lethal activity of microbial inactivation by chlorine (Al-Qadiri et al., 2008).

The bacteria cell membrane which is external to the cell wall protects the cell from the environment and contains compounds such as phospholipids and lipopolysaccharides which are stabilized by Ca^{2+} and Mg^{2+}. On the cytoplasmic membrane, molecules of disinfectants may infiltrate the cytoplasmic membrane either by active diffusion which is non-specific and slow or by more specific active transport which enables products to build up in bacteria after transformation or binding to a membrane protein. Additionally, hypochlorous acid is reported to be highly potent against constituents of cell membrane required for cell division, and it affects metabolic processes by inhibiting synthesis of proteins required for DNA replication and cell division (McKenna and Davies, 1988). Disruption of the cell membrane produces a bacteriostatic effect and causes severe damage to the cell integrity. This results in the inhibition of the cellular metabolic activity due to the effect on the respiratory enzymes and transport systems localized near the cell surface. For instance, Haas et al. (1980) recorded an increase in the level of high molecular weight total extractable Total Organic Carbon and high total extractable UV-absorbing compounds of chorine-treated cells after discharge into the bacterial suspension compared to the untreated bacterial cells. Disruption of the cell membrane indicates a bacteriostatic action by chlorine on the cell membrane.

It has been hypothesized that the carbonaceous materials present on the cell membrane and cell wall are targets for oxidation by free chlorine which increases the chlorine demand at the membrane-chlorine interphase. This leads to alteration of the cell wall and cell membrane causing a discharge of intracellular materials in the form of nitrogenous materials and increase in the formation of chloramines and DNA destruction (Shang and Blatchley, 2001).

Another hypothesis describes a reduced respiratory activity of chlorine inhibition on certain enzymes in the glucose respiratory pathway. This is a major mechanism of bacterial inactivation by chlorine (Knox et al., 1948). For example, tricarboxylic acid cycle enzymes including malate synthase (GlcB) and malate dehydrogenase (Mdh) and enzymes involved in glycolytic pathway, glyceraldehyde 3'-phosphate dehydrogenase and phosphoglycerate kinase controlling the cell metabolism, have been oxidized by HOCl. Further details on HOClmediated oxidation of the respiratory pathway can be found in the review by Gray et al. (2013).

The lethal effect of chlorine is partly due to the production of hydroxyl radicals (HO·) generated by a Fenton-type reaction during oxidation of Fe-S clusters that results in the release of cellular iron (Candeias et al., 1994). Fe-S cluster assembly and genes associated with cysteine biosynthesis are targets for oxidative damage (Wang et al., 2009). HOCl causes damage to the primary and secondary amines of nucleotide bases resulting in DNA breakage (Grey et al. 2013).

HOCl rapidly reacts with purine and pyrimidine bases of nucleotides, organosulphur or aromatic moieties of amino acids, and the suphhydryl moiety of glutathione (Deborde and Guten, 2008). HOCl reacts poorly with the amide groups of amino acids that bind to peptides but exhibits high reactivity with the N-terminal amino acids in peptidoglycan and proteins, or the aromatic or amine side chains of tyrosine, tryptophan, histidine, cysteine, methionine, or the free organosuphur, or lysine groups in proteins (Sharma et al., 2013). Oxidation of cysteine residues form sulphenic, sulphinic and sulphonic acids, while methionine is oxidized to methionine sulphoxide (Met-SO) or methionine sulphone (Davies, 2005).

8.5 MICROBIAL RESISTANCE TO CHLORINE DISINFECTION IN WATER TREATMENT

The survival of chlorine resistant pathogens is an impairment to water purification often resulting in public health implications of infectious diseases including giardiasis, gastroenteritis, diarrhoea and typhoid (Okoh et al., 2010). For example, pathogens, including *Escherichia coli*, *Faecal enterococci*, *Clostridium perfringens*, *Salmonella*, *Shigella dysenteriae*, *Vibrio* species and coliphages, exhibiting high resistance to chlorine, have been recovered even at free chlorine residuals of 0.32 mg/L in final effluents of wastewater in various communities in South Africa (Odjadjare and Olaniran, 2015; Teklehaimanot et al., 2014). This indicates that the wastewater final effluent discharged is not in compliance with the South African regulatory standard of 0 CFU/100 mL bacterial counts.

Salmonella species can enter into viable but nonculturable state in the presence of stressors including disinfectants, afterwards returning to actively metabolizing state to potentiate infections (Oliver, 2000). Also, enterococci are usually found in humans and animal faeces, and faecal enterococci, e.g. *E. faecalis*, *E. faecium*, *E. durans* and *E. hirae*, have been identified as indicators of sewage contamination (Castillo-Rojas et al., 2013). Resistance of *Enterococcus faecalis* to antimicrobial agents such as chlorine has been documented in water/wastewater treatment systems (Iweriebor et al., 2015). Also, vancomycin resistant enterococci was detected in four wastewater treatment plants and in reclaimed treated wastewater systems in the United States (Carey et al., 2016; Rosenberg-Goldstein et al., 2014). Mycobacteria are highly resistant to chlorination. Five species of mycobacteria were recovered from treated water samples collected from 25 water treatment plants in the United States with two species, *M. mucogenicum* and *M. phocacium*, identified as emerging pathogenic mycobacteria (King et al., 2016).

Two serotypes of Adenovirus, AdV40 and AdV41, documented as etiological agents of human enteric diseases have been detected in wastewater final effluents in our previous report (Osuolale and Okoh, 2015) as well as in chlorinated water systems (Carducci and Verani, 2013). Human adenoviruses (HAdV) are predominantly found in environmental water samples such as drinking water (Katukiza et al., 2014) and are capable of long-term survival in water at low temperatures (Carducci and Verani, 2013). HAdV have exhibited high resistance in disinfection processes with 2.3 log_{10} and 3.3 log_{10} reductions in the viral load observed after 60 minutes (De Abreu et al., 2012) and were also detected in two wastewater treatment plants (El-Senousy et al., 2013).

Cryptosporidium parvum and *Giardia lamblia* are able to resist conventional disinfection processes including chlorine and are persistent in final effluents of wastewater (Dungeni and Momba, 2010). Their infectivity is enhanced by presence of an outer shell of oocyst and cysts that protect them from environmental stress and oxidants such as chlorine.

The reactivation of *E. coli* and *Salmonella* species in water distributed from water reclamation plants (WRP) and distribution networks after treatment with ozone and chlorine in china, were ascribed to maintenance of membrane integrity and reverse transcriptional activity of the bacteria (Lin et al., 2016). Low chlorine dosage and high nutrient levels supported the regrowth of microbial pathogens.

8.5.1 Exogenous Factors Implicated in Pathogen Response to Chlorine Disinfection

Some exogenous factors have been implicated as key contributors to microbial response to chlorine disinfectant in water treatment (DWAF, 1996). Some of the exogenous factors include temperature, pH, attachment to surfaces, turbidity, organic matter, climate change, chlorine dose and contact time (Gomes et al., 2016; Coronel-Olivares et al., 2011)

Chlorine is effective at high temperatures ranging between 0° and 25°C and pH 7–9 for Ct_{99} inactivation of microbial pathogens depending on the chlorine dose and contact time (WHO, 2011). At low pH values, the efficiency of the active moiety of chlorine, HOCl, is increased. At pH of 4.0, the active species, HOCl dominates and has a greater activity compared to hypochlorite ion which is predominant at higher pH values (LeChevallier and Au, 2004).

Organic matter and suspended particles increase the turbidity of water and substantially increase chlorine demand. Turbidity is a good predictor of total coliform counts and heterotrophic plate counts (Weishaar et al., 2003), and the correlation between these parameters designates the persistence of bacteria coliforms in chlorinated waters, with turbidity values between 3.8 and 84 NTU (Reilly and Kippin, 1983). Several accounts have demonstrated the correlation of high turbidity levels with high frequency of microbial survival in chlorinated water/wastewater. For example, investigations carried out on final effluents of wastewater treatment plants in the Eastern Cape Province, South Africa, revealed a high frequency of *Vibrio* spp. (24%–40% abundance) owing primarily to a high turbidity level of 2.16–17.5 NTU among other factors considered (Igbinosa et al., 2011). A similar study showed final effluents of high turbidity and containing elevated concentration organic matter accounted for the high frequency of multiple antibiotics resistant (MAR) *Pseudomonas* spp. in counts ranging from 1.08×10^4 to 2.66×10^4 CFU/mL (Odjadjare et al., 2012). Also, high counts of *Escherichia coli* and *Vibrio* ranging from 0.6×10^3 to 0.86×10^4 CFU/mL and 1.44×10^3 to 1.86×10^4 CFU/mL for respectively correlated high turbidity levels of 4.02–43.20 NTU in our previous report (Osuolale and Okoh, 2015).

Chlorine is less efficient in the presence of organic matter and suspended particles which serve as a protective barrier against the oxidative influence of chlorine and promotes the resistance of cell envelopes in chlorinated effluents. The concentration of organic solutes at a solid surface may increase free chlorine demand at the interface, and this decreases the availability of chlorine and weakens the disinfection process (Virto et al., 2005). Suspended particles increase attachment of microorganism to surfaces providing an enabling environment for biofilm formation. Chlorine radicals are impenetrable to biofilms. Furthermore, high organic matter contents are considered as sources of precursors of DBPs in water disinfection (Gomes et al., 2016).

8.5.2 Intrinsic Factors Mediating Chlorine Resistance in Microbial Pathogens

Physiologically, bacteria protect themselves against reactive oxygen species (ROS) or oxidative stress (Georgopoulos, 2006) by activating molecular chaperones or heat shock proteins that protect the newly synthesized polypeptide chains and assemble subunits from aggregating into non-functional structures. These chaperones include the GroEL/ES system or the DnaK/DnaK/DnaJ/GrpE system (Georgopoulos, 2006) which is ATP- and nucleotide-dependent (Winter et al., 2005) and winter Hsp33 which induces the formation of DTT-resistant Hsp33 oligomers in the presence of HOCl (Bajszár and Dekonenko, 2010) and matrix metalloproteinases (MMP-7, MMP-8 and MM-9) (Fu et al., 2004). The *dps* gene encodes the DNA-binding protein at the stationary phase and oxyR regulon in the exponential phase. All *recA* mutants of *E. coli* were susceptible to lethal doses of HOCl above 0.6 mg/L, while at sublethal doses of <0.6mg/L, *dps*rec A mutants were protected from oxidative damage (Fu et al., 2001). The high amounts of aggregated proteins found in bacteria depleted of molecular chaperone Hsp33 activated by HOCl increased the sensitivity of bacteria to bleach than the wild-type bacteria (Gray et al., 2013).

Biofilm formation is another physiological response of bacteria to specific environmental stress. Microorganisms protect themselves from stressors such as disinfectants, UV light, pH, moisture, heat fluctuations and viruses through the formation of biofilms (Winter et al., 2005).

Well-defined bacteria responses under HOCl stress conditions include defences against oxidative stress, protein unfolding and sulphur starvation. Resistance against oxidants such as chlorine, peracetic acid and hydrogen peroxide has been activated by inducing oxyR and soxRS regulon and by production of catalase to reduce damages resulting from production of radicals (Fu et al., 2001). The dodecameric transcriptional factor, hypT (YjiE) shields *E. coli* from hypochlorous acid by upregulating genes regulating cysteine, methionine and sulphur metabolism and the downregulation of iron acquisition and homeostasis genes (Krige et al., 2009).

The cell membrane acts as a permeability barrier to the penetration of biocide, thereby limiting the concentration of the biocide that will be absorbed into the cell.

Alteration in cell membrane permeability reduces the penetration of chlorine into the cytoplasm. The mechanism of alteration is controlled by the balance of porin proteins and the diffusion channels acting as channels for small hydrophilic compounds (Morente et al., 2013). The cell envelope is a site for chlorine reactive species, and Gram negative bacteria such as *E. coli* O157: H7 respond to chlorine treatment by activating the gene ycfR or *ycfQ* gene which increases the hydrophobicity of the outer membrane (Deng et al., 2011).

Cross-resistance and co-resistance mechanisms occur in which chlorine facilitates the transfer of conjugative genes through exchange of genetic materials between antibiotic resistant and antibiotic sensitive bacteria strains (Xu et al., 2016). Cross-resistance was observed between antibiotics resistant and chorine resistant strains of *A. baumanii* (10^9 CFU/mL) in bacterial suspensions, exposed to 0–4 mg/L chlorine doses at 30, 60, 90 and 120 seconds (Karumathil et al., 2014). Co-resistance was demonstrated with the presence of β-lactam resistant genes in filtered water (39.2%), drinking water (34.9%) and treated water (76.5%). Chlorine facilitated multidrug and aminoglycoside resistances in drinking water and plasmid replication as revealed by metagenomic analysis (Shi et al., 2013). It was further deduced that chorine could influence antibiotics resistance through alteration of membrane permeability and inhibition of molecular transport. Drinking water showed abundance of antibiotic resistant *E. coli* (63%), *Enterobacter* (18.5%), *Klebsiella* (16.7%) and *Acinetobacter* (1.9%) in filtered water (FW), chlorinated water samples and samples collected from clear tank of Beihekou Tap water plant in Nanjing, China (Shi et al., 2013).

Mechanisms of pathogen co-resistance to chlorine and antibiotic resistance included selective survival of resistant bacteria population at sublethal doses of chlorine and prior exposure to chlorine resulting in selective pressure against further oxidative stress (Peeters et al., 2010).

8.6 CHLORINE RESISTANT PATHOGENS VERSUS DBPs IN WATER TREATMENT: PRIORITIZING PUBLIC HEALTH CONCERNS

Approximately 3.4 million people, mostly children, die annually from waterborne infections (WHO, 2002) and approximately 748 million people do not have access to clean sources of drinking water (WHO, 2015). The availability and accessibility to safe water is evidence to sustainable development (United Nations, 2015). In Sub-Saharan Africa, waterborne disease outbreaks are annual occurrences. Cholera has become endemic in some parts of Nigeria and occurs frequently as a pandemic in other regions in Nigeria (NCDC, 2019).

Water quality regulations are becoming more stringent on production of quality water and effluent (Rietveld et al., 2009) as treated effluents are finding more uses as reclamation water in agriculture and water consumption. More importantly, the incidence of antibiotic resistant bacteria (ARBs) and antibiotic resistant genes (ARGs) in chlorinated effluents is becoming a growing concern to public health. The emergence of chlorine resistance in waterborne pathogens including *Cryptosporidium parvum*, *Giardia lamblia*, *Helicobacter pylori*, *E. coli* O157:H7, multidrug resistant *Mycobacterium tuberculosis*, Rotavirus, Human papilloma virus, Norovirus, *Legionella pneumophila*, *Campylobacter* sp. and severe acute respiratory syndrome

(SARS) Coronavirus will reduce the efficacy of prophylaxis and antibiotics therapy (Zhang et al., 2009). Bacteria pathogens such as *E. coli* O157:H7, *Vibrio*, *Salmonella* spp. and multidrug resistant *Mycobacterium tuberculosis* in source water are potential risks of infections such as diarrhoea, dysentery, typhoid, skin and tissue infections and gastrointestinal diseases (Igbinosa et al., 2010).

Wastewater effluents are often hotspots for chlorine resistant pathogens containing an array of pathogenic microorganisms, and they require adequate disinfection to prevent discharge of resistant pathogens. Sublethal dosing could result in developing resistance through prior exposure of pathogens to disinfectant and through mutations. Sublethally injured bacteria could recover through their repair mechanisms and channelled by effluent into waterbodies. Increase in chlorine concentration may provide a sustainable solution to chlorine resistance due to the advantages of leaving a high residual chlorine in water distribution systems, effluents discharge and reclamation waters. In our previous studies, we reported that *E. coli*, *Citrobacter* and *Enterococcus* species were inactivated at free chlorine doses of 1.5, 1.0 and 1.0. mg/L compared to poor removal efficiency of bacteria at the recommended chlorine residual of 0.5 mg/L (Owoseni and Okoh, 2017a, b; Owoseni et al., 2017).

In vitro and in vivo studies have evidenced the likely ecological and human health risks resulting from exposure to DBPs when chlorine reacts with organic matter precursors. These include dissolved organic matter (DOM) in wastewater, humic and fulvic acids in drinking water, pharmaceuticals, pesticides and dyes in reclaimed water (Pan et al., 2017; Postigo and Richardson, 2014; Richardson 2011). The mutagenicity and genotoxicity of DBPs such as THMs, HAA, N-nitrosodimethylamine (NDMA) and chloroforms have been associated with increased risks of reproductive and developmental health and bladder cancer (Smith et al., 2016; Grelier et al., 2015; Zamyadi et al., 2012). DBPs are toxicity drivers highlighting future pathways for epidemiological studies.

In spite of the evidence documented on DBPs, many studies have shown inconsistent and inconclusive epidemiological data which do not reflect the association between DBPs and reproductive health (Nieuwenhuijsen et al., 2000). Using the quantitative application of in vitro cytotoxicity and genotoxicity assays based on Chinese hamster ovary (CHO) cells, greater than 100 DBPs has been generated (Sisti et al., 1998). However, the CHO cells were shown to be deprived of certain metabolic features such as expression of the enzyme glutathione S-transferase (GST) theta-1 (GSTT1) that are essential for activation of DBPs to mutagens (Li and William, 2018). In addition, no statistically significant correlations have been found between drinking water consumption containing high concentrations of THM4 and occurrence of bladder cancer in many meta-analysis of epidemiological studies (Li and William, 2018). Most DBPs including THM4 concentrations have been hypothesized to cause cancers due to their carcinogenic nature and not necessarily because they are the principal drivers of cancer risk (Li and William, 2018). WHO (2003) showed inconclusive reports on the asymptomatic (higher than 100 mg/L chlorine dose) and symptomatic effects of high chlorine doses (1.9–100 mg/L chlorine dose) causing various histopathological infections in rats. Proposed factors to be considered in integrating DBPs into epidemiological studies include the consistent concentration measurements, spatiotemporal variations in concentrations and pharmacokinetics of emerging DBPs (Li and William, 2018).

To achieve an optimized water treatment process, attention should be given to all treatment procedures to minimize the concentration of organic matter precursors leading to the formation of DBPs. For instance, removal of organic precursors through the pretreatment process such as coagulation resulted in DOC removal by 2%–85% and by biofiltration process thus reducing oxidant demand (Du et al., 2017; Li and William, 2018). Characterizing the mechanisms of pathogen inactivation can also facilitate better understanding of how chemical disinfectants including chlorine transform various biomolecules of pathogens (Li and William, 2018). Chorine remains a widely used disinfectant in the 21st century relevant to water treatment efficiency as it may leave a disinfectant residual, offering some level of disinfection capacity as well as a real-time measurable marker of microbial demand for chlorine (Hrudey and Hrudey 2007).

8.7 CONCLUSIONS AND FUTURE PERSPECTIVES

Chlorination of water serves as a mainstay for public health safety providing cost-effective disinfection, simple technology and easy access globally and documented pathogen reduction efficiency, particularly in low- and middle-income countries. Chlorine is a potent oxidant that attacks cell biomolecules and causes protein aggregation. In spite of water treatment guidelines to control pathogen survival and ensure water quality, microbial pathogens are surviving in chlorinated water through various resistant processes. Chlorine resistance is aided by extrinsic and intrinsic factors which reduces disinfection efficiency. The inactivation of various microbial groups, particularly the protozoans, in the recommended chlorine dose of 15 mg/min/L at 30 minutes needs to be considered to ensure an effective chlorination process. The knowledge of DBPs formation in chlorinated water as drivers of cancer and risks to reproductive and developmental health can spur more investigations into the mechanisms of pathogen inactivation by chlorine which will increase pathogen inactivation and chlorine efficiency. These various perspectives will provide more data that will spur the review of existing guidelines and ensure optimization of chlorination in water treatment.

REFERENCES

Al-Adhami, B. H., Nichols, R. A. B., Kusel, J. R. , O'Grady, J. and Smith, H. V. (2007). Detection of UV-induced thymine dimers in individual Cryptosporidium parvum and Cryptosporidium hominis oocytes by immunoflourescence microscopy. *Appl Environ Microbiol.* 73: 947–55.

Al-Qadiri, H. M., Al-Alami, N. I., Al-Holy, M. A. and Rasco, B. A. (2008). Using Fourier transform infrared (FT-IR) absorbance spectroscopy and multivariate analysis to study the effect of chlorine-induced bacterial injury in water. *J Agric Food Chem.* 8;56(19):8992–7. doi: 10.1021/jf801604p.

Arnold, B. F. and Coleford Jr.J. M. (2007). Treating water with chlorine at point-of-use to improve water quality and reduce child diarrhea in developing countries: a systematic review and meta-analysis. *Am J Trop Hyg.* 76(2): 354–64.

Bajszár, G. and Dekonenko, A. (2010). Stress-induced Hsp70 gene expression and inactivation of *Cryptosporidium parvum* oocysts by chlorine-based oxidants stress-induced Hsp70 gene expression and inactivation of *Cryptosporidium parvum* oocysts by chlorine-based oxidants. *Appl Environ Microbiol.* 76(6):1732–9. doi: 10.1128/AEM. 02353-09.

Baker, M. N. and Taras, M. J. (1981). *The Quest for Pure Water: A History of the Twentieth Century, Volume 1 and 2.* AWWA, Denver.

Bergeron, S., Boopathy, R., Nathaniel, R., Corbin, A. and LaFleur, G. (2015) Presence of antibiotic resistant bacteria and antibiotic resistance genes in raw source water and treated drinking water. *Int Biodeterior Biodegradation.* 102:370–4. doi: 10.1016/j.ibiod.2015.04.017.

Bouki. C., Venieri, D. and Diamadopoulos, E. (2013). Detection and fate of antibiotic resistant bacteria in wastewater treatment plants : a review. *Ecotoxicol Environ Saf.* 91:1–9.

Candeias, L. P., Stratford, M. R. and Wardman, P. (1994) Formation of hydroxyl radicals on reaction of hypochlorous acid with ferrocyanide, a model iron(II) complex. *Free Radic Res.* 20:241–9.

Carducci, A. and Verani, M. (2013) Effects of bacterial, chemical, physical and meteorological variables on virus removal by a wastewater treatment plant. *Food Environ Virol.* 5(1):69–76.

Carey, S. A., Goldstein, R. E. R., Gibbs, S. G., Claye, E., He, X. and Sapkota, A.R. (2016). Occurrence of vancomycin-resistant and -susceptible *Enterococcus* spp. in reclaimed water used for spray irrigation. *Environ Res.* 147:350–5.

Castillo-Rojas, G., Mazari-Hiríart, M., Ponce de León, S., Amieva-Fernández, R. I., Agis-Juárez, R. A., Huebner, J., et al. (2013). Comparison of *Enterococcus faecium* and *Enterococcus faecalis* strains isolated from water and clinical samples: antimicrobial susceptibility and genetic relationships. *PLoS One.* 8(4):e59491.

Centre for Disease Control and Prevention. (1999). Outbreak of *Vibrio parahaemolyticus* infection associated with eating raw oysters and clams harvested from long island Sound- Connecticut, New Jersey, and New York. *Morb Mortal Wkly Rep.* 48:48–51.

Centre for Disease Control and Prevention. (2012). *Effect of chlorination on inactivating selected pathogen.* Centers for Disease Control and Prevention, National Centre for Emerging and Zoonotic Infectious Diseases (NCEZID), Division of foodborne, Waterborne, and Environmental Diseases (DFWED). https://www.cdc.gov/safewater/effectiveness-on-pathogens.html. Accessed on 24th April, 2020.

Chevremont, A. C., Boudenne, J. L., Coulomb, B. and Farnet, A. M. (2013). Impact of watering with UV- LED-treated wastewater on microbial and physico-chemical parameters of soil. *Water Res.* 47 (6):1971–82.

Cho, M., Kim, J., Kim, J. Y., Yoon, J. and Kim, J. H. (2010). Mechanisms of *Escherichia coli* inactivation by several disinfectants. *Water Res.* 44 (11):3410–8.

Coronel-Olivares, C., Reyes-Gómez, L. M., Hernández-Muñoz, A., Martínez-Falcón, A. P., Vázquez-Rodríguez, G. A. and Iturbe, U. (2011). Chlorine disinfection of *Pseudomonas aeruginosa*, total coliforms, *Escherichia coli* and *Enterococcus faecalis*: revisiting reclaimed water regulations. *Water Sci Technol.* 64(11):2151–7.

Council directive 98/83/EC (1998). Offi. Off J Eur Communities.

Crump, J. A., Otieno, P. O., Slutsker, L., Keswick, B. H., Rosen, D. H., Hoekstra, R. M., Vulule, J. M. and Luby, S.P. (2005). Household based treatment of drinking water with flocculant-disinfectant for preventing diarrhoea in areas with turbid source water in rural western Kenya: cluster randomised controlled trial. *BMJ* 331: 478.

Davies, M. J. (2005). The oxidative environment and protein damage. *Biochim Biophys Acta.* 1703(2):93–109.

Deborde, M. and von Gunten, U. (2008) Reactions of chlorine with inorganic and organic compounds during water treatment-kinetics and mechanisms: a critical review. *Water Res.* 42(1–2):13–51.

De Abreu Corrêa, A., Carratala, A., Barardi, C. R. M., Calvo, M., Girones, R. and Bofill-Mas, S. (2012). Comparative inactivation of murine norovirus, human adenovirus, and human JC polyomavirus by chlorine in seawater. *Appl Environ Microbiol.*78(18):6450–7.

Deng, K., Wang, S., Rui, X., Zhang, W. and Tortorello, M. L. (2011). Functional analysis of ycfR and ycfQ in *Escherichia coli* O157:H7 linked to outbreaks of illness associated with fresh produce. *Appl Environ Microbiol.* 77 (12): 3952–59.

Department of Water Affiars. (1996). *South African Water Quality Guidelines*, Volume 1, Domestic Water Use, 2nd edition. Pretoria, South Africa.

Department of Water Affiars. (2013). *National Water Act.* Department of Water Affairs, Pretoria, South Africa, pp. 3–32.

Drazic, A., Gebendorfer. K. M., Mak, S., Steiner, A., Krause, M., Bepperling, A. and Winter, J. (2014). Tetramers are the activation-competent species of the HOCl-specific transcription factor HypT. *J Biol Chem.* 289(2):977–86. doi:10.1074/jbc.M113.521401.

Du, Y., Lv, X., Wu, Q., Zhang, D., Zhou, Y., Peng., L. and Hu, H. (2017). Formation and control of disinfection byproducts and toxicity during reclaimed water chlorination. A review. *J. Environ. Sci.* doi: 10.1016/j.jes.2017.01.013.

Dungeni, M. and Momba, M. N. B (2010). The abundance of *Cryptosporidium* and *Giardia* spp. in treated effluents produced by four wastewater treatment plants in the Gauteng Province of South Africa. *Water SA*. 36 (4):425–31.

Dungeni, M., Van der Merwe, R. R., and Momba, M. N. B (2010). Abundance of pathogenic bacteria and viral indicators in chlorinated effluents produced by four wastewater treatment plants in the Gauteng Province, South Africa. *Water Res.* 36(5):607–14.

El-Senousy, W. M., Barakat, A. B., Ghanem, H. E. and Kamel, M. A. (2013). Molecular epidemiology of human adenoviruses and rotaviruses as candidate viral indicators in the Egyptian sewage and water samples. *World Appl Scences J.* 27(10):1235–47.

Environmental Protection Agency Victoria (2002). *Disinfection of Treated Wastewater I Guidelines for Environmental Management.* Environmental Protection Agency, Victoria, Australia. Available at: https://www.epa.vic.gov.au/about-epa/publications/730. Accessed on 24th April, 2020.

Environmental Protection Agency. (2011). *Water Treatment Manual: Disinfection.* Environmental Protection Agency, Wexford, Ireland.

Finnegan, M., Linley, E., Denyer, S. P., McDonnell, G., Simons, C. and Maillard, J. Y. (2010). Mode of action of hydrogen peroxide and other oxidizing agents: differences between liquid and gas forms. *J Antimicrob Chemother.* 65(10):2108–15.

Fu, X., Kao, J. L. F., Bergt, C., Kassim, S. Y., Huq, N. P., d'Avignon, A., et al. (2004). Oxidative cross-linking of tryptophan to glycine restrains matrix metalloproteinase activity: specific structural motifs control protein oxidation. *J Biol Chem.* 279(8):6209–12.

Fu, X., Kassim, S. Y., Parks, W. C., and Heinecke, J. W. (2001). Hypochlorous acid oxygenates the cysteine switch domain of pro-matrilysin (MMP-7): a mechanism for matrix metalloproteinase activation and atherosclerotic pla- que rupture by myeloperoxidase. *J. Biol. Chem.* 276:41279–41287.

GB 5749-2006. (2006). China GB5749-2006 drinking water guidelines.pdf. pp. 1–17.

Georgopoulos, C. (2006). Toothpicks, serendipity and the emergence of the *Escherichia coli* DnaK (Hsp70) and GroEL (Hsp60) chaperone machines. *Genetics.* 174:1699–707.

Gomes, I. B., Simões, M. and Simões, L. C. (2016). The effects of sodium hypochlorite against selected drinking water-isolated bacteria in planktonic and sessile states. *Sci Total Environ.* 565:40–8 doi: 10.1016/j.scitotenv.2016.04.136 PMID: 27156214.

Gray, M. J., Wholey, W. Y. and Jakob, U. (2013). Bacterial responses to reactive chlorine species. *Annu Rev Microbiol.* 67:141–60. doi:10.1146/annurev-micro-102912-1425.

Green P. A., Vorosmarty, C. J., Harrison, I., Farrell, T., Saenz, L. and Feteke B. M. (2015). Freshwater ecosystem services supporting humans: Pivoting from water crisis to water solutions. *Glob Environ Chang.* 34:108–18 doi: 10.1016/j.gloenvcha.2015.06.007.

Grellier, J., Rushton, L., Briggs, D. J. and Nieuwenhuijsen, M. J. (2015). Assessing the human health impacts of exposure to disinfection by-products – a critical review of concepts and methods. *Environ. Int.* 78:61–81, doi: 10.1016/j. envint.2015.02.003.

Haas, C. N., Engelbrecht, R. S., Haas, C. N. and York, N. (1980). Physiological alterations of vegetative microorganisms. *Water Environ Fed.* 52(7):1976–89.

HDR Engineering Inc. (2001). *Handbook of Public Water Systems*, 2nd ed., John Wiley & Sons Inc, New York.

Helbling, D. E. and Vanbriesen, J. M. (2007). Free chlorine demand and cell survival of microbial suspensions. *Water Res.* 41(19):4424–34. Available from: http://www.ncbi.nlm.nih.gov/pubmed/17624396.

Hrudey, S. E. and Hrudey, E. J. (2007). Published case studies of waterborne disease outbreaks-Evidence of a recurrent threat. *Water Environ. Res.* 79:233–45.

Igbinosa, E. O, Obi, C. L. and Okoh, A. I. (2011). Seasonal abundance and distribution of Vibrio species in the treated effluent of wastewater treatment facilities in suburban and urban communities of Eastern Cape Province, South Africa. *J Microbiol.* 49(2):224–32.

Igbinosa, E. O. and Okoh, A. I. (2009). Impact of discharge wastewater effluents on the physico-chemical qualities of a receiving watershed in a typical rural community. *Int J Environ Sci Technol.* 6(2):175–82. doi: 10.1007/BF03327619.

Igbinosa, E. O. and Okoh, A. I. (2010). *Vibrio fluvialis*: An unusual enteric pathogen of increasing public health concern. *Int. Environ. Res. Pub. Healt.* 7:3628–43.

Iweriebor, B. C., Gaqavu, S., Obi, L. C., Nwodo, U. U. and Okoh, A. I. (2015). Antibiotic susceptibilities of *Enterococcus* species isolated from hospital and domestic wastewater effluents in Alice, eastern cape province of South Africa. *Int J Environ Res Public Health.* 12(4):4231–46.

Karumathil, D. P., Yin, H. B. Kollanoor-Johny, A. and Venkitanarayanan, K. (2014). Effect of chlorine exposure on the survival and antibiotic gene expression of multidrug resistant Acinetobacter baumannii in water. *Int J Environ Res Public Health.* 11(2):1844–54. doi: 10.3390/ijerph110201844 PMID:24514427.

Katukiza, A. Y., Ronteltap, M., van der Steen, P., Foppen, J.W. A. and Lens, P.N.L. (2014) Quantification of microbial risks to human health caused by waterborne viruses and bacteria in an urban slum. *J Appl Microbiol.* 116(2):447–63.

King, D.N., Donohue, M. J., Vesper, S. J., Villegas, E. N., Ware, M. W., Vogel, M. E., et al. (2016). Microbial pathogens in source and treated waters from drinking water treatment plants in the United States and implications for human health. *Sci Total Environ.* 562:987–95.

Knox, W. E., Stumpf, P. K., Green, D. E. and Auerbach, V. H. (1948). The inhibition of sulfhydryl enzymes as the basis. *J Bacteriol.* 55:451–8.

Koivunen, J. and Heinonen-Tanski, H. (2005). Inactivation of enteric microorganisms with chemical disinfectants, UV irradiation and combined chemical/UV treatments. *Water Res.* 39(8):1519–26.

Krige, M. (2009). Molecular investigation of the chlorine and antibiotic resistance mechanisms of *Escherichia coli* isolated from natural water sources in the western Cape Marilyn Krige *Thesis Presented in Partial Fulfilment of the Requirements for the Degree of Master of S. Stellenbosch University*, pp. 1–214.

Laubusch, E. J. (1964). Chlorination and other chlorine institute. *Disinfect Proces.* 3/4:331–42.

Lazarova, V., Savoye, P., Janex, M.L., Blatchley, E.R. I. and Pommepuy, M. (1999). Advanced wastewater disinfection technologies: State of the art and perspectives. *Water Sci Technol.* 40:203–13.

Leal, J. L. (1909). The Sterilization Plant of the Jersey City Water Supply at Boonton, N.J. *Proceedings of American Water works Association*, pp. 100–109.

Lechevaillier, M.W. and Au, K.K. (2004). *Water Treatment and Pathogen Control, Process Efficiency in Achieving Safe Drinking Water.* International Water Association Publishing, London, UK, pp. 1–112.

Leichert, L., Gehrke, F., Gudiseva, H. V., Blackwell, T., Ilbert, M., Walker, A. K., et al. (2007). Quantifying changes in the thiol redox proteome upon oxidaive stress in vivo. *PNAS*. 105(24):8197–202. doi: 10.1073 pnas.0707723105.

Li, X. and William, M. (2018). Drinking water disinfection by-products (DBPs) and human health effects: Multidisciplinary challenges and opportunities. *Environ. Sci. Technol.*, 52:1681–1689.

Lin, Y., Li, D., Zeng, S. and He, M. (2016). Changes of microbial composition during wastewater reclamation and distribution systems revealed by high- throughput sequencing analyses. *Front Environ Sci Eng*. 10(3):539–47. doi:10.1007/s11783-016-0830-5.

Martínez-Hernández, S., Vázquez-Rodríguez, G. A., Beltrán-Hernández, R. I., Prieto-García, F., Miranda-López, J. M., Franco-Abuín, C. M., Álvarez-Hernández, Iturbe, U. and Coronel-Olivares, C. (2013). Resistance and inactivation kinetics of bacterial strains isolated from the non-chlorinated and chlorinated effluents of a WWTP. *Int J Environ Res Public Health*, 10(8):3363–83.doi:10.3390/ijerph10083363.

Martino R. R. (2003). Water, an essential element for life. *A Contribution of the Delegation of the Holy See on the Occasion of the Third World Water Forum*, Kyoto, Japan.

McKenna, S. M. and Davies, J. A. (1988). The inhibition of bacterial growth by hypochlorous acid possible role in the bactericidal activity of phagocytes. *Biochem J*. 254:685–92.

Michael, I., Rizzo, L., McArdell, C. S., Manaia, C. M., Merlin, C., Schwartz, T., et al. (2013). Urban wastewater treatment plants as hotspots for the release of antibiotics in the environment: a review. *Water Res*. 47(3):957–95.

Morente, E. O., Fernández-Fuentes, M. A. G. Burgos, M. J., Abriouel, H., Pulido, R. P. and Gálvez, A. (2013). Biocide tolerance in bacteria. *Int J Food Microbiol*. 162(1):13–25. doi: 10.1016/j.ijfoodmicro.2012.12.028.

Nieuwenhuijsen, M. J., Toledano, M. B., Eaton, N. E., Fawell, J. and Elliott, P. (2000). Chlorination disinfection byproducts in water and their association with adverse reproductive outcomes : a review. *Occup. Environ. Med.* 57, 73–85.

Nigeria Centre for Disease Control. (2019). An update of cholera outbreak in Nigeria. Situation report. www.ncdc.gov.ng. (accessed 07 May, 2020).

Nigerian Industrial Standard. (2007). Nigerian standard for Drinking Water Quality. Standards organization of Nigeria. ICS 13. 060. 20. pp 1–22.

Odjadjare, E. C. and Olaniran, A. O. (2015). Prevalence of antimicrobial resistant and virulent *Salmonella* spp. in treated effluent and receiving aquatic milieu of wastewater treatment plants in Durban, South Africa. *Int J Environ Res Public Health*. 12(8):9692–713. doi: 10.3390/ijerph9062092 PMID: 22829792.

Odjadjare, E. E., Igbinosa, E. O. Mordi, R., Igere, B., Igeleke, C. L. and Okoh, A. I. (2012). Prevalence of multiple antibiotics resistant (MAR) *Pseudomonas* species in the final effluents of three municipal wastewater treatment facilities in South Africa. *Int J Environ Res Public Health*. 9(6):2092–107.

Okoh, A. I, Sibanda, T. and Gusha, S. S. (2010). Inadequately treated wastewater as a source of human enteric viruses in the environment. *Int J Environ Res Pub Health*. 2620–37. doi: 10.3390/ijerph7103628.

Oliver, J. D. (2000). *Public Health Significance of Viable but Nonculturable Bacteria. Non-Culturable Microorganisms in the Environment* (eds R. R. Colwell and D. J. Grimes) American Society for Microbiology Press, Washington, DC.

Osuolale, O. and Okoh, A. (2015) Assessment of the physicochemical qualities and Prevalence of *Escherichia coli* and *Vibrios* in the final effluents of two wastewater treatment plants in South Africa: ecological and public health implications. *Int J Environ Res Public Health*. 12(10):13399–412. doi: 10.3390/ijerph10083363PMID:23924881.

Owoseni, M. and Okoh, A. (2017a) Evidence of emerging challenge of chlorine tolerance of *Enterococcus* species recovered from wastewater treatment plants. *Int. Biodeterior. Biodegradation* 120: 216–223. doi: 10.1016/j.ibiod.2017.02.016.

Owoseni, M. and Okoh, I. (2017b). Assessment of chlorine tolerance profile of Citrobacter species recovered from wastewater treatment plants in Eastern Cape, South Africa. *Environ. Monit. Assess.* 189 (201): 1–12. doi: 10.1007/s10661-017-5900-z.

Owoseni, M. C., Olaniran, A. O. and Okoh, A. I. (2017). Chlorine tolerance and inactivation of *Escherichia coli* recovered from Wastewater treatment plants in the Eastern Cape, South Africa. *Appl. Sci.* 810; doi: 10.3390/app7080810.

Pan, Y., Wang, Y., Li, A., Xu, B., Xian, Q., Shuang, C., Shi, P. and Zhou, Q. (2017). Detection, formation and occurrence of 13 new polar phenolic chlorinated and brominated disinfection byproducts in drinking water. *Water Res.* doi: 10.1016/j.watres.2017.01.037.

Peeters, E., Sass, A., Mahenthiralingam, E., Nelis, H. and Coenye, T. (2010). Transcriptional response of *Burkholderia cenocepacia* J2315 sessile cells to treatments with high doses of hydrogen peroxide and sodium hypochlorite. *BMC Genomics.* 11(90): 1–18.

Postigo, C. and Richardson, S. D. (2014). Transformation of pharmaceuticals during oxidation/disinfection processes in drinking water treatment. *J. Hazard. Mater.* 279: 461–475. doi: 10.1016/j.jhazmat.2014.07.029.

Reilly, J. K., Kippin, J. (1983). Relationship of bacterial counts with turbidity and free chlorine in two distribution systems. *J Am Water Work Assoc.*75(6):309–12.

Richardson, S. D. (2011). Disinfection By-products: formation and occurrence of drinking water. In: Nriagu, J. O. (Ed.), *The Encyclopedia of Environmental Health* vol. 2. Elsevier, Burlington, pp. 110–136.

Rietveld, L. C., Meijer, L., Smeets, P. W. M. H. and Van Der, J.P. (2009). Assessment of Cryptosporidium in wastewater reuse for drinking water purposes : a case study for the city of Amsterdam, The Netherlands. *Water SA.* 35(2):211–5.

Rosen, H., Klebanoff, S. J., Wang, Y., Brot, N., Heinecke, J. W. and Fu, X. (2009). Methionine oxidation contributes to bacterial killing by the myeloperoxidase system of neutrophils. *PNAS.* 106 (44). doi: www.pnas.org/cgi/doi/10.1073.

Rosenberg-Goldstein, R. E., Micallef, S. A., Gibbs, S. G., George, A., Claye, E., Sapkota, A., et al. (2014). Detection of vancomycin-resistant enterococci (VRE) at four U.S. wastewater treatment plants that provide effluent for reuse. *Sci Total Environ.*466–467:404–11.

Shang, C. and Blatchley III, E. R (2001). Chlorination of pure bacterial cultures in aqueous solution. *Pergamon.* 35(1):244–54.

Shannon, K. E., Lee, D. Y., Trevors, J.T. and Beaudette, L. A. (2007). Application of real-time quantitative PCR for the detection of selected bacterial pathogens during municipal wastewater treatment. *Sci Total Environ.*382 (1):121–9.

Sharma, V. K., Liu, F., Tolan, S., Sohn, M., Kim, H. and Oturan, M. A. (2013). Oxidation of β-lactam antibiotics by ferrate(VI). *Chem Eng J.* 221:446–51.

Shi, P., Jia, S., Zhang, X., Zhang, T., Cheng, S. and Li, A. (2013). Metagenomic insights into chlorination effects on microbial antibiotic resistance in drinking water. *Water Res.* 47 (1):111–120.

Sisti, M., Albano, A. and Brandi, G. (1998). Bactericidal effect of chlorine on motile Aeromonas spp. in drinking water supplies and influence of temperature on disinfection efficacy. *Lett Appl Microbiol Appl Microbiol.* 26:347–51.

Smith, R. B., Edwards, S. C., Best, N., Wright, J., Nieuwenhuijsen, M. J. and Toledano, M. B. (2016). Research | Children's health birth weight, ethnicity, and exposure to trihalomethanes and haloacetic acids in drinking water during pregnancy in the born in Bradford cohort. *Environ. Health Perspect.* 124:681–689.

Sonune, A., and Ghate, R. (2004). Development in wastewater treatment methods. *Desalination* 167:55–63.

Stopforth, J. D., Mai, T., Kottapalli, B. and Samadpour, M. (2008). Effect of acidified sodium chlorite, chlorine, and acidified electrolyzed water on *Escherichia coli* 0157:H7, Salmonella and Listeria monocytogenes inoculated onto leafy greens. *J Food Prot.* 71(3): 625–8.

Teklehaimanot, G. Z., Coetzee, M. A. A. and Momba, M. N. B. (2014). Faecal pollution loads in the wastewater effluents and receiving water bodies: a potential threat to the health of Sedibeng and Soshanguve communities, South Africa. *Environ Sci Pollut Res Int.* 21(16):9589–603.

Tiernan, M. F. (1948). Controlling the green goddess. *J Am Work Assoc.* 40(10): 1042–50.
United Nations (2010). *The Millennium development goals Report.* United Nations, New York.
United Nations. (2015). *The Critical Role of Water in Achieving the Sustainable Development Goals : Synthesis of Knowledge and Recommendations for Effective Framing, Monitoring, and Capacity Development.* United Nations, New York.
Virto, R., Mañas, P., Álvarez, I., Condon, S. and Raso, J. (2005). Membrane damage and microbial inactivation by chlorine in the absence and presence of a chlorine-demanding substrate. *Appl Environ Microbiol.* 71(9): 5022–8.
Wang, S., Deng, K., Zaremba, S., Deng, X., Lin. C., Wang, Q, Tortorello, M. L. and Zhang, W. (2009). Transcriptomic response of *Escherichia coli* O157:H7 to oxidative stress. Appl Environ Microbiol. 75(19):6110–23. doi: 10.1128/AEM.00914-09.
Weishaar, J. L., Aiken, G.R., Bergamaschi, B.A., Fram, M.S., Fujii, R. and Mopper, K. (2003). Evaluation of specific ultraviolet absorbance as an indicator of the chemical composition and reactivity of dissolved organic carbon. *Environ Sci Technol.* 37:4702–8.
White, G. C. (1999). *Handbook of Chlorination.* John Wiley & Sons Inc, New York.
WHO. (2003). Chlorine in Drinking-water. Background document for the development of WHO guidelines for drinking-water Quality. http://www.who.int/water_sanitation_health/dwq/chlorine.pdf.
WHO. (2006). Guidelines for drinking-water quality. *World Heal Organ.* 1:1–597.
WHO. (2011). Guidelines for drinking-water quality. *Guidel Drink Qual.* 1:104–8 Available on: http://www.ncbi.nlm.nih.gov/pubmed/15806952.
WHO. (2015). *World Health Statistics,* pp. 1–161.
Winter, J., Ilbert, M., Graf, P.C.F., Ozcelik, D. and Jakob, U. (2008). Bleach activates a redox-regulated chaperone by oxidative protein unfolding. *Cell.* 135(4):691–701.
Xu, L., Ouyang, W., Qian, Y., Su, C., Su, J. and Chen, H. (2016). High-throughput profiling of antibiotic resistance genes in drinking water treatment plants and distribution systems. *Environ Pollut.* 213:119–26. doi: 10.1016/j.envpol.2016.02.013.
Ye, X. Y., Ming, X., Zhang, Y. L., Xiao, W. Q., Huang, X. N., Cao, Y. G. and Gu, K. D. (2012). Real-time PCR detection of enteric viruses in source water and treated drinking water in Wuhan, China. *Curr Microbiol.* 65 (3):244–53. doi: 10.1007/s00284-012-0152-1.
Yodder, J. S., Hlavsa, M. C., Craun, G. F., Hil, V., Roberts, V., Yu, P. A., Hicks, L. A., Alexander, N. T., Calderon, R. L., Roy, S. L. and Beach, M. J. (2008). *Surveillance for Waterborne Disease and Outbreaks Associated with Recreational Water Use and Other Aquatic facility-assocaited health events-United States,* 2005-2006, CDC MMWR, 57 (SS09)., pp. 1–29. US Government Printing Office (GPO), Washington, DC. Available on: http://www.cdc.gov/mmwr/preview/mmwrhtml/ss5709a1.htm.
Zamyadi, A., Ho, L., Newcombe, G., Bustamante, H. and Prévost, M. (2012). Fate of toxic cyanobacterial cells and disinfection by-products formation after chlorination. *Water Res.* 46:1524–1535.
Zhang, X. X., Zhang, T. and Fang, H. P. H. (2009). Antibiotic resistance genes in water environment. *Appl Microbiol Biotechnol.* 82:397–414.

Index

Note: **Bold** page numbers refer to tables and *italic* page numbers refer to figures.